Innovation, learning, and technological dynamism of developing countries

Innovation, learning, and technological dynamism of developing countries

Edited by Sunil Mani and Henny Romijn

TOKYO • NEW YORK • PARIS

© United Nations University, 2004

The views expressed in this publication are those of the authors and do not necessarily reflect the views of the United Nations University.

United Nations University Press
United Nations University, 53-70, Jingumae 5-chome,
Shibuya-ku, Tokyo, 150-8925, Japan
Tel: +81-3-3499-2811   Fax: +81-3-3406-7345
E-mail: sales@hq.unu.edu   general enquiries: press@hq.unu.edu
http://www.unu.edu

United Nations University Office at the United Nations, New York
2 United Nations Plaza, Room DC2-2062, New York, NY 10017, USA
Tel: +1-212-963-6387   Fax: +1-212-371-9454
E-mail: unuona@ony.unu.edu

United Nations University Press is the publishing division of the United Nations University.

Cover design by Sese-Paul Design

Printed in Hong Kong

UNUP-1097
ISBN 92-808-1097-9

Library of Congress Cataloging-in-Publication Data

Innovation, learning, and technological dynamism of developing countries / edited by Sunil Mani and Henny Romijn.
   p.   cm.
Includes bibliographical references and index.
ISBN 9280810979 (pbk.)
1. Technological innovations—Developing countries. I. Mani, Sunil.
II. Romijn, Henny.
T173.8.I552   2004
338′.064′091724–dc22   2004007422

Contents

Acknowledgements vii

Contributors viii

List of figures ix

List of tables xi

Abbreviations xiii

Preface xvi

1 Introduction 1
  Sunil Mani and Henny Romijn

2 Exports of high technology products from developing countries:
  Are the figures real or are they statistical artefacts? 12
  Sunil Mani

3 Development strategies and innovation policies in globalisation:
  The case of Singapore 48
  Alexander Ebner

v

4 Evolution of the civil aircraft manufacturing system of
  innovation: A case study in Brazil 77
  Rosane Argou Marques

5 The political economy of technology policy: The automotive
  sector in Brazil (1950–2000)..................................... 107
  Effie Kesidou

6 Technological learning in small-enterprise clusters: Conceptual
  framework and policy implications.............................. 135
  Marjolein C. J. Caniëls and Henny Romijn

7 The contribution of skilled workers in the diffusion of
  knowledge in the Philippines..................................... 158
  Niels Beerepoot

8 Understanding growth dynamism and its constraints in high
  technology clusters in developing countries: A study of
  Bangalore, southern India... 178
  M. Vijayabaskar and Girija Krishnaswamy

9 Culture, innovation, and economic development: The case of
  the South Indian ICT clusters 202
  Florian Arun Taeube

Index 229

Acknowledgements

The papers included in this book were initially presented at a conference at the United Nations University–Institute for New Technologies (UNU-INTECH) at Maastricht. The Dutch Research School for Resource Studies for Development (CERES) and the European Association for Development Research and Training Institutes (EADI) both sponsored the conference. Our grateful thanks go to these two organisations for their support and encouragement throughout the course of this project. A number of our colleagues offered comments and criticisms on the papers, especially when they were presented at the above conference. We are particularly grateful for suggestions, comments, or other forms of assistance received from Louk de la Rive Box, Lynn Mytelka, Peter Knorringa, Keith Smith, and Bart Verspagen. Further discussions with colleagues at the working group on Science and Technology for Development of the EADI and the comments made by the two anonymous referees have greatly enhanced the quality of the manuscript.

Eveline in de Braek and Marijke Roolvink-Batty, our secretaries at INTECH, deserve special thanks for their timely help in getting the manuscript ready for publication.

Finally both of us would like to thank our respective home institutions, UNU-INTECH and the Eindhoven Centre for Innovation Studies at Eindhoven University of Technology, for facilitating this project.

Sunil Mani
Henny Romijn

Contributors

Niels Beerepoot, AGIDS, University of Amsterdam, Amsterdam, the Netherlands.

Marjolein Caniëls, The Open University of the Netherlands, Heerlen, the Netherlands.

Alexander Ebner, University of Erfurt, Erfurt, Germany.

Effie Kesidou, Eindhoven Centre for Innovation Studies, Technische Universiteit Eindhoven, Eindhoven, the Netherlands.

Girija Krishnaswamy, Australian Catholic University, Sydney, Australia.

Sunil Mani, United Nations University/Institute for New Technologies, Maastricht, the Netherlands.

Rosane Argou Marques, Science and Technology Policy Research, University of Sussex, Sussex, UK.

Henny Romijn, Eindhoven Centre for Innovation Studies, Technische Universiteit Eindhoven, Eindhoven, the Netherlands.

Florian Arun Taeube, Johann Wolfgang Goethe-University, Frankfurt, Germany.

M. Vijayabaskar, Madras Institute of Development Studies, Chennai, India.

List of figures

1.1	Book structure	3
2.1	Share of developing countries in world exports of manufactured products	14
2.2	Ratio of World Bank estimates to INTECH estimates	20
2.3	Trends in high technology content of world exports of manufactured products	23
2.4	The increasing share of developing countries in total world exports of high tech exports, 1988–1998	24
2.5	The catching up of developing countries with respect to high tech exports (at the aggregate level)	26
2.6	The catching up of developing countries by product level	27
2.7	The catching up of specific countries with the United States	28
2.8	Competitiveness of developed and developing countries in high tech exports	29
2.9	Net export ratio of developed and developing countries, 1988–1998	30
2.10	The top countries in terms of average net export ratio, 1968–1998	30
2.11	US patent activity by inventor country and grant year, 1985–1998	36
2.12	Trends in R&D expenditure and in research intensity in Singapore	42

4.1	Knowledge and production systems in the civil aircraft manufacturing system of innovation	83
4.2	Knowledge and production systems in the Brazilian civil aircraft manufacturing system of innovation	96
5.1	Total vehicle production and exports of the automotive industry	127
5.2	Total revenue of auto parts manufacturers	128
5.3	Auto parts industry output: Percentage distribution by destination	129
6.1	Integrating macro with meso	138
9.1	Regional distribution of interviewees in India	217
9.2	State-wise distribution of interviewees in South and West India	217
9.3	Social background of interviewees in India	218
9.4	Ethnicity of interview partners in India	218

List of tables

2.1	Structure of exports of developing countries, 1988 and 1998	15
2.2	Value of exports of manufactured products from developing countries, 1991–1997	16
2.3	High tech products list by the OECD	19
2.4	The leading high tech exporters in the world, 1997	21
2.5	Comparison of World Bank and INTECH datasets	22
2.6	Concentration in exports of high tech products	25
2.7	Ranking of developed and developing countries according to high tech export intensity	25
2.8	RCA indices of leading high tech exporters from the developing world	29
2.9	Structure of high tech exports of developing and developed countries	32
2.10	Specialisation of developed and developing countries within electronics and office and computing equipment, 1991 and 1997	33
2.11	The top 10 high tech exports by developing and developed countries	35
2.12	Top 15 most common US patent classes for inventors from Korea, Singapore, and Malaysia	38
2.13	Indicators of technological competitiveness	39
2.14	Model of Singaporean electronics technology development	44
3.1	Output and national income in Singapore, 1965–2001	52

3.2	Structural change in the Singapore economy, 1965–1995 ...	54
3.3	Foreign and local net investment commitments in Singapore's manufacturing industries, 1972–1999	55
3.4	Innovation input indicators in Singapore, 1981–1999	64
3.5	Expenditure on R&D by research area, 1999................	65
3.6	R&D expenditures of local and foreign companies by industry, 1999...	66
4.1	The evolution of aircraft models produced in Brazil	91
4.2	Economic performance of the Brazilian aircraft manufacturing sector	92
4.3	Embraer economic indicators, 1990–2002...................	92
5.1	Contrasting approaches to technology policy (TP): A summary	114
6.1	Direct effects of agglomeration advantages on the technological efforts (TE) of the firm	141
7.1	Segmentation of the labour market in the furniture industry in Cebu...	171
8.1	VC/angel investments in high tech firms in India............	193
9.1	Number of engineering colleges and enrolment	213
9.2	Ethnicity of interviewees in the Indian software industry and their share in the national population	218

Abbreviations

ASEAN	Association of South East Asian Nations
BASI	innovation system of the aircraft industry
BEFIEX	Fiscal Incentives for Exports in Brazil
BNDE	Banco Nacional de Desenvolvimento Económico (Brazil)
CE	collective efficiency
CERES	Dutch Research School for Resource Studies for Development
CMM	Capability Maturity Model
COMTRADE	UN Commodity Trade Statistics Database
CTA	Aerospace Technical Centre (Brazil)
EADI	European Association for Development Research and Training Institutes
EDB	Economic Development Board (Singapore)
EU	European Union
FDI	foreign direct investment
GDP	gross domestic product
GEIA	Grupo Executivo da Indústria Automobilística (Brazil)
GERD	gross expenditure on R&D
GNP	gross national product
IC	integrated circuit

ICT	information and communication technology
IFI	Institute for Development and Coordination of the Aerospace Industry (Brazil)
IISc	Indian Institute of Science
IITs	Indian Institutes of Technology
IMF	International Monetary Fund
INTECH	Institute for New Technologies
IT	information technology
ITA	Technological Institute of Aeronautics (Brazil)
KS	knowledge system
LDCs	less developed countries
MIT	Massachusetts Institute of Technology
MN	multinational
MNCs	multinational corporations
NASA	National Aeronautics & Space Administration (USA)
NICs	newly industrialised countries
NRI	non-resident Indian
NSTB	National Science and Technology Board
NTU	Nanyang Technological University
NUS	National University of Singapore
NYSE	New York Stock Exchange
OECD	Organisation for Economic Co-operation and Development
PCB	printed circuit board
PS	production system
RCA	revealed comparative advantage
RSEs	research scientists and engineers
S&T	science and technology
SEI	Software Engineering Institute (USA)
SI	systems of innovation
SIC	Standard Industrial Classification
SITC	Standard International Trade Classification
SMEs	small- and medium-sized enterprises
SPRING	Singapore Standards, Productivity and Innovation Board
STP	software technology park
TC	technological capability
TDICI	The Technology Development and Information Co. of India Ltd.
TI	Texas Instruments
TNC	transnational corporations
TPAC	Technology Policy Assessment Center

USPTO	United States Patents and Trademarks Office
VC	venture capital
WB	World Bank
WTO	World Trade Organization

Preface

Sunil Mani and Henny Romijn

The association of the term technological dynamism with that of developing countries does indeed raise some eyebrows. This is because most developing countries (the definition of which is based on a certain predetermined threshold of average per capita GDP) continue to employ and reproduce technologies which are generated elsewhere. However, this situation is slowly but steadily changing, although academic research on the issue of technological dynamism has been restricted to a few countries from the East Asian regions variously described as Asian tigers and cubs.

Much of the discussion about the technological dynamism of developing countries has used a narrow neoclassical economic framework. In this set-up, technical progress is very often quantified using a summary measure such as the rate of growth of total factor productivity. The recent debate within the context of East Asian economies has shown that this debate has more to do with quirks of methodology and indeed the type of dataset that is employed to compute this indicator of the contribution made by technology. Admittedly the discussions along these lines have become very technical and therefore have run the risk of failing to identify specific instances of technological dynamism. In contrast, the papers included in this volume employ a refreshingly new framework in identifying cases of technological dynamism across a range of countries and industries. The cases discussed vary from the recent growth of the computer software industry in India to the case of the aircraft industry in

Brazil. The cases covered thus touch upon technological dynamism observed in both manufacturing and service oriented industries. Another related issue that has merited attention in the book is the contribution of a specific organisational form, namely the geographic agglomeration of firms engaged in the production of related and complementary items – or clustering, as it is more popularly known – in making sectors more dynamic from the technological point of view.

1
Introduction

Sunil Mani and Henny Romijn

Firms in developing countries secure their requisite technology through two sources: foreign and domestic. Assimilation of technologies created elsewhere is still the dominant mechanism for most of these countries. However, some developing countries recently have attained the coveted status of creators of new technologies. This status has been achieved through a process of learning and incremental innovation, facilitated by the introduction of complementary policies, institutions, and organisational arrangements.

Yet, in most discussions on technology development in the context of developing countries, the term "developing countries" is still being used as if referring to a homogeneous group having an average per capita income below a certain predetermined threshold (with the possible exception of the unique experience of a handful of countries from East Asia, which have been singled out as special). Such a formalistic way of grouping developing countries together tends to overlook the increasing dissimilarities between those that have, at least, the potential to catch up and even create new technologies on their own, and those that do not have this capacity.

Against the background of increasing differentiation in technological performance, the aim of this book is to highlight notable instances of successful technological catch-up and endogenous technology creation by developing countries and to point to important factors behind such success. The book starts by mapping out the broad patterns of recent technological dynamism among developing countries (chap. 2). Then it goes on to explore specific cases of technological dynamism, and concentrates on important factors that have contributed to this success. Two broad themes are explored. The first theme, which takes centre stage in chapters 3–5, relates to the role played by the policy environment and the systemic properties of the innovation process. These issues provide the setting for a detailed study of some notable technological successes in specific developing countries. The second theme, explored in chapters 6–9, deals with a particular organisational strategy that is increasingly being resorted to by policy makers from the South to enhance the innovative behaviour of firms and institutions. This organisational method draws its sustenance from agglomeration economies and is popularly referred to as "clustering". The technological dynamism of specific clusters is analysed in these later chapters and the economic and non-economic factors that have contributed to this are identified. The main research issues addressed here are: how clustering of firms producing similar products or services can lead to improved innovative performance of the firms constituting the cluster; and under which conditions this is likely to occur. Contrasting experiences from low tech clusters from Pakistan and the Philippines and from a high tech cluster from India shed light on these questions. A refreshingly new line of analysis is the role of culture and other non-economic variables in explaining cluster dynamics.

There is thus a symbiotic link between the different chapters in this book (fig. 1.1). The first chapter (chap. 2) sets the scene for the following chapters, by empirically identifying a set of countries that have shown some spectacular results in terms of local technology development. It is found that developing countries are increasingly becoming exporters of high technology or technology-intensive products, thus reflecting the general trend observed in world trade. This chapter begins by asking the question whether this is a statistical artefact or whether there are real instances of technology dynamism. The chapters that follow furnish evidence supporting the latter view, through a further exploration of specific instances of countries, industries, and local regions in which such technological dynamism has been noticed. They contain an in-depth examination of the significance of technological performance in particular cases, and delve into the issue of where that dynamism originated and how it emerged. The chapters employ a variety of analytical approaches deriving mainly from economics, geography, and anthropology, thus reflecting the diverse disciplinary backgrounds and training of the authors. This gives each chapter its own flavour, and adds to the richness of the book as a whole. This is thus a unique collection of chapters focusing on technology and development, areas relatively unexplored in the literature.

We now turn our attention to individual chapters.

INTRODUCTION 3

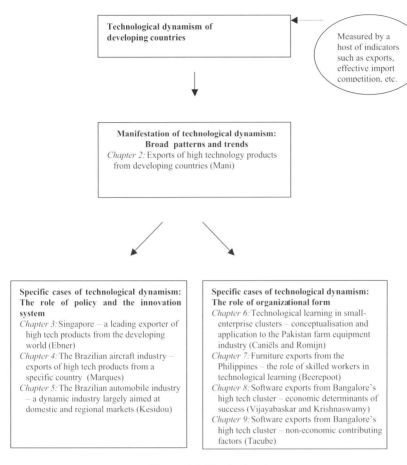

Figure 1.1 Book structure

One important way to measure the technological dynamism of developing countries is by observing the structure of their exports.[1] If the structure of exports is changing towards seemingly sophisticated items, then one may conclude that these countries are becoming technologically dynamic. Chapter 2 by Mani subjects this proposition to systematic empirical scrutiny. This chapter first develops consistent time-series data on the exports of high tech products from developing countries. Analysis of the data shows that developing countries are increasingly becoming exporters of manufactured products compared with primary products as in the past. Second, world trade is increasingly becoming a trade in high tech products. What is more striking is the significant increase in the technology content of exports from developing countries: very nearly a quarter of the exports from developing countries are now high tech products. Third, the share of developing countries in high tech exports has dramatically increased from about 8 per cent in 1988 to about 23 per cent in 1997. However, this increase is largely concentrated in a few countries. The chapter then seeks to explain whether or not these developing countries are real exporters of high tech products. This is accomplished by a careful examination of the degree of product specialisation by both developed and developing countries, by examining their record with respect to patenting, and finally by analysing certain indicators of high tech competitiveness. The technological capability of even the high tech exporting countries varies greatly. At one end of the spectrum lie Korea and, increasingly, Singapore, which have the local capability to design, manufacture, and export high tech items. Malaysia is somewhere in the middle, while Thailand and the Philippines appear to be at the other end with (relatively speaking) a very low capability. However, the group of developing countries as a whole is fast catching up with the developed countries, so it may be unwise to write off this performance as a mere statistical artefact.

The three chapters that follow document specific cases of technological dynamism in developing countries, and shed detailed light on supporting policies and the systemic features of innovation systems. Singapore is one of the leading high tech exporters of the developing world. Ebner (chap. 3) analyses the nature and extent of technological development in Singapore and traces the role of innovation policies in explaining its technological success. He begins by mapping out the spectacular technological performance of the country. The aggregate research intensity, for instance, rose significantly from about 0.26 per cent in the early 1980s to close to 2 per cent by the late 1980s. This performance can be traced to a number of proactive state policies. To illustrate, formal state involvement in supporting domestic technology creation commenced with the establishment of the National Science and Technology Board (known since 2002 as the Agency for Science, Technology and Research). In addition, a large number public research institutes were created and several fiscal incentives were put in place. Technical education was also promoted in a significant manner so that the country had an adequate supply of trained personnel. At the same time the country also pursued a vigorous policy of promoting foreign companies, particularly as regards manufacturing. The main argument of the chapter is that the general relevance of the Singaporean case lies in the increasing importance of local agglomerations of high value-added activities which are fuelled by the process of globalisation, for they receive their structural impact from their roles as strategic hubs in multinational enterprise networks.

Brazil occupies a unique position among the developing countries. Although no more than 9 per cent of its manufactured exports are high tech products, it is the only country from the developing world to have a civil aircraft manufacturing industry. Aircraft, such as the ERJ 145 50-seater jet and the ERJ 135 37-seater jet, are manufactured by Embraer, a formerly state-owned aerospace organisation which was privatised in 1994. Marques (chap. 4) analyses the innovation system of the aircraft industry (referred to as the BASI) especially after its privatisation. Embraer, which employs almost 12,000 people and also manufactures a number of components, has particular strength in constructing undercarriages, and is a subcontractor for various overseas manufacturers of fixed/rotary wing aircraft. Brazil does not build civil aircraft jet engines, so companies such as Rolls-Royce and General Electric have established local subsidiaries to fulfil this need. Marques shows that two sets of changes have affected the BASI after privatisation in the mid-1990s. The first set of changes affected the production system resulting in: (i) increased participation of Embraer in the world civil aircraft market; (ii) increased participation of first tier foreign suppliers in product development; (iii) localisation of subsidiaries of first tier foreign suppliers in Brazil; and (iv) expanded possibilities for local second tier firms supplying first tier foreign suppliers. The second set of changes is related to the knowledge system and includes: (i) a decrease in the role of the Aerospace Technical Centre (CTA) as one of the main actors in the national knowledge system; (ii) an increase in the role of foreign suppliers and foreign science and technology (S&T) institutes in transferring technological knowledge to the Brazilian civil aircraft production system; (iii) an increase in the role of other Brazilian organisations (universities and technological centres) in supporting the application of basic technological knowledge developed abroad to the local production system needs; and (iv) evolution of the knowledge and production systems towards a wider configuration post-1994. Other sources (in addition to Embraer and the CTA) of technological knowledge for the local suppliers have become available now that there are localising foreign first tier suppliers closely related to local suppliers as well as other S&T institutes (national and foreign).

Kesidou (chap. 5) analyses the role of technology policy in making the Brazilian automotive industry very dynamic. Brazil's automotive industry is the tenth largest in the world and contributes about 12 per cent of GDP. Brazil plays host to the largest number of car assembly plants in the world and is the second largest target of foreign direct investment. In the 1990s vehicle production increased over 60 per cent, domestic car sales grew by 65 per cent, and exports expanded by around 9 per cent. In 1997, at its peak, Brazil produced more than 2 million vehicles and earned almost US$5 billion from exports; motor vehicles accounted for almost 10 per cent of the total value of Brazil's exports for that year. By the turn of the century production had dropped somewhat because of the economic situation, but Brazil will soon be one of the five largest automotive producers in the world, with the most modern factories. Brazil exports vehicles to other members of Mercosur. This chapter analyses historically the role of technology policy in shaping the industry since its inception in 1952. As regards the nature and extent of innovation policy, two phases are evident: the first phase concerns technology policy with respect to the automotive industry from 1952 until 1974, while the second phase relates to a structural shift in policy which is discernible during the post-1975 period. Although the state was very much involved in technology policy during the first phase, other agents, such as auto parts manufacturers and transnational corporations (TNCs), also contributed directly and indirectly to the decision-making. In contrast, the second phase is characterised by a gradual withdrawal of the state from technology policy, leaving the decision-making largely in the hands of the TNCs. However, Kesidou argues that, although the power of the federal state has been reduced, there has been considerable decentralisation of decision-making with respect to economic matters away from the federal government to governments in the states (provinces) and to cities within states. An interesting aspect of the second phase, and especially the period since the mid-1990s, is the fact that car makers' investment strategies in Brazil were basically of the "market seeking with moderate trade effort" (in the Dunning sense) variety. In other words, the industry took advantage of liberalisation to import more without altering its export orientation beyond perhaps including Mercosur in its market access strategy. So, while the industry has shown much dynamism (in the narrower sense), it has also become much more externally dependent.

Both the Singaporean and Brazilian cases thus show that developing countries can achieve technological maturity, and that they can also progress from being incremental innovators to being real innovators, provided the governments of these countries have clearly articulated policies for shaping the national and sectoral systems of innovation. However, in both countries increasing integration of their economies with the rest of the world has meant that the domestic systems of innovation have come under some strain. While Singapore has successfully managed to combine a programme of systematic globalisation with the strengthening of its domestic system of innovation, the same is not the case with Brazil, if judged by the experiences of the BASI and the auto industry.

While the importance of supportive policies for the promotion of well-functioning systems for domestic technological learning and innovation has been recognised for some time among researchers and policy makers, awareness of the importance of spatial factors in this process is more recent. Approximately from the 1990s onwards, a voluminous body of literature has emerged which highlights the importance of regional agglomeration for economic growth and competitiveness, in both advanced and developing countries. However, despite liberal use of terms such as "learning region", "regional innovation system", and "innovative milieu", so far little systematic attention has been devoted to the *technological* factors that supposedly underpin these economic benefits. This important emerging gap in the literature is taken up in the four remaining chapters of this book.

Caniëls and Romijn (chap. 6) set the scene with a framework that shows how regional clustering of companies engaged in similar and complementary economic activities could conceivably foster technological dynamism at the company level. Taking Marshall's agglomeration benefits as their starting point, they elaborate a taxonomy of mechanisms through which these benefits may impact on firms' investments in technological effort and on the effectiveness of their learning processes. The authors then proceed to apply the framework to the farm equipment industry in the Punjab region of Pakistan, a notable case of effective import competition among eight local agglomerations of small- and medium-sized firms. The clusters, which emerged in the 1950s, expanded to employ approximately 5,000 workers by the 1990s and supported phenomenal productivity growth in agriculture during the "green revolution". Although by the mid-1990s it could still be described as a low tech industry, it was by then capable of manufacturing and incrementally adapting well over 50 items of farm equipment. Caniëls and Romijn illustrate how the clustered nature of the industry facilitated the process of reverse engineering that led to this achievement. Their analysis describes a range of spontaneous effects, including the establishment of specialised parts and components suppliers due to the viable size of the local market; transaction cost advantages and local knowledge spillovers arising from face-to-face interactions with suppliers and customers; and local dissemination of production knowledge and skills through inter-firm movement of workers. Several policy-relevant insights emerge, which deviate from those of earlier cluster studies with a less explicit focus on technological dynamism. The latter have linked cluster dynamism mainly to effective local inter-firm cooperation, and have accordingly tended to emphasise the importance of building trust among local parties. Instead, Caniëls and Romijn conclude that it might be better to single out a few dynamic producers for help with their innovation efforts, and let spontaneous non-collaborative clustering effects take care of wider diffusion. However, their analysis also suggests that such supply-side support policies cannot be a substitute for the need to address stagnant economic environments in which producers have no incentives to innovate. Agglomeration can facilitate innovation, but it is by no means a sufficient condition for it to occur.

One of the agglomeration mechanisms identified by Caniëls and Romijn, namely the contribution of skilled workers to the diffusion of skills and knowledge, is chosen for detailed study by Beerepoot in his chapter about a furniture cluster in the Philippines (chap. 7). Workers are frequently mentioned as a major carrier of the region-specific knowledge by which local communities gain and retain competitive advantage. However, very little is actually known about how they transmit knowledge, and who is important for what purpose. The importance of local labour market spillovers can hardly be overemphasised in export-based clusters in developing countries, many of which consist of agglomerations of labour-intensive industries manufacturing, for example, clothes, footwear, furniture, and toys. Their competitive advantage is subject to quick erosion due to rising local labour costs and the recent entry of cheaper competitors (China!) on the world trade scene. As Beerepoot points out, under such conditions it is pertinent to identify the feasibility of pursuing an industrial strategy based on localised learning in order to cope with these challenges. The effective utilisation of an industry's best asset, a highly skilled flexible workforce, is the greatest challenge in this development trajectory. Beerepoot investigated this issue in Cebu, where a furniture cluster that had its origin in craft production grew into the prime furniture exporter in Southeast Asia during the 1970s and 1980s. The spectacular dynamism of the industry is evident from the fact that there are around 175 exporters in Cebu who provide employment to roughly 45,000 workers and jointly contribute about 1.8 per cent of Philippine GDP. However, since the 1990s the industry has ceded market share, mainly to China and Indonesia. Beerepoot found that keen competition pits parties against each other and makes them protective of their knowledge. The patterns are basically the same across the stages in the value chain and horizontally between exporters. For example, entrepreneurs increasingly resort to informal subcontracting as a means to cut costs and risk, not as a platform for sharing new ideas. Knowledge-protecting behaviour is most in evidence among entrepreneurs, managers, and designers, who stand to lose most from sharing their specialised trade contacts and internal production knowledge. Skilled production workers, in contrast, will readily teach their traditional craft skills to others on the job. On the whole, Beerepoot's findings suggest that embarking on a collective learning strategy to cope with the cluster's competitiveness problems will not be easy in the prevailing competitive environment. Indirectly, his research also concurs with the previous chapter about the limited scope for interventions aimed at nurturing social institutions such as trust.

The final two chapters focus on the recent rise of India as a major export platform for software services since the early 1990s: exports accounted for US$5,000 million in 2001. Bangalore, the largest cluster and earliest starter, hosts several hundreds of software firms, including around 70 leading multinationals such as Motorola, Texas Instruments, and Hewlett-Packard. Bangalore also provided growing space for Indian-owned companies, which have now become famous multinationals in their own right. Firms are beginning to climb the value chain, moving on from the provision of simple service tasks, such as data processing and operation of call centres, to offering integrated service packages and software development. Not surprisingly, this spectacular growth of technological dynamism has recently begun to attract widespread international attention. Much effort has been devoted to explaining the success in terms of particular constellations of resource endowments and Bangalore's position in the international division of labour. The last two chapters in the book add to this emerging body of literature by assessing the role played by other factors which have remained relatively underexplored to date.

Vijayabaskar and Krishnaswamy's chapter (chap. 8) draws attention to several organisational mechanisms specific to the Bangalore cluster, which may critically influence its ability to upgrade technologically, and thus its capacity to sustain its competitiveness in the world market. A particularly striking observation is that internal inter-firm specialisation in the cluster is still very limited. The authors ascribe this in part to the fact that most firms are still involved in highly similar labour-intensive operations. Few have entered into the domain of more skill-intensive operations such as design. This is in fact very difficult to do, since the prospective users are located far away, and close ongoing contact is often needed to cater effectively for their needs. This organisational feature is thus rooted in Bangalore's particular position in the international division of labour. Another notable characteristic of the cluster is the scarcity of local inter-firm networks. Networks tend to be international rather than local: there is no culture of local knowledge sharing, as in Silicon Valley. Again this can be related to Bangalore's position as a supplier of software solutions to major overseas clients, who often enforce secrecy on their software suppliers and discourage local outsourcing by them. Further features unfavourable to technological upgrading are identified in the labour market. In the absence of sufficient local opportunities for upward mobility, workers with the best skills tend to migrate abroad (although some later return with more experience). Labour mobility problems are particularly bad for local small- and medium-scale firms, which cannot offer attractive enough career options. At this developmental stage of the cluster, the costs of high labour market spillovers appear to outweigh the benefits. While not discounting the considerable benefits from export-led development that have occurred in the cluster, the authors caution that the cluster's current excessive export dependence may well be the most important constraint on its further development. They argue that state policies should be directed towards achieving a better domestic–foreign balance.

The second chapter about Indian software by Taeube (chap. 9) investigates the role played by cultural factors in the recent achievements of the information and communication technology (ICT) sector in southern (and to some extent western) India. Taeube asks whether there is more behind this success than the constellation of economic and geographic features highlighted in the literature so far. Awareness of possible cultural influences is especially vital when possibilities for replication of India's ICT "model" are being contemplated, which is always tempting for policy makers. After surveying how different writers have conceptualised the links between culture and economic development, Taeube proceeds to make the concept operational in the specific context of South India. Hinduism serves as the relevant frame of reference here, not in the narrow sense of religion, but in the broader sense of Indian civilisation pervading all spheres of life. Two particular aspects, namely caste and ethnicity, are singled out as important attributes that could have a bearing on values and behaviour conducive to success specifically in the ICT sector. A preliminary analysis of data from surveys previously conducted in the sector supports the importance of these cultural attributes. In particular, the sector enjoys the participation of a disproportionate number of *Brahmins*, the social group in India that has been traditionally found in occupations concerned with knowledge, learning, and teaching. At the same time, the *Vaishyas*, the merchant and trader castes, are underrepresented in the sector. Further, most of the key people in the Indian software industry are located in South India and come from a South Indian background in terms of ethnicity or family affiliation. Taeube then tries to unravel the ways in which this caste and ethnic profile is suited to the type of work that the ICT sector offers, and supplies some interesting matches. While his work is still in its preliminary stages, these early findings are promising and suggest that there is more behind India's recent ICT successes than "hard" economic and geographic determinants.

In sum, this book adds to the very small but growing literature on the phenomenon of the technological dynamism of developing countries (for a recent paper on this theme, see Mahmood and Singh, 2003). In the existing methodology, this dynamism has been measured through the application of standard innovation outcome indicators such as patent statistics. The book begins by marshalling considerable quantitative evidence on inter-country variations in technological dynamism. However, the uniqueness of the book lies not only in its extension of the analysis from standard macro aggregates across countries to specific instances of dynamism, but also in the incorporation of non-economic variables, such as culture and ethnicity, in explaining this dynamism. This detailed and interdisciplinary approach renders our analysis of this growing phenomenon quite distinct.

Note

1. The ability to manufacture and sell in the international market is taken as a good indicator of a country's competitiveness.

REFERENCE

Mahmood, P. I. and J. Singh (2003) "Technological Dynamism in Asia", *Research Policy* 32(6): 1031–1054.

2
Exports of high technology products from developing countries: Are the figures real or are they statistical artefacts?

Sunil Mani

1 Introduction

In the ever burgeoning literature on economic growth, one of the most important issues is the growth performance of a handful of East Asian countries. These countries have earned a reputation, despite the recent financial crisis, for being one of the fastest growing regions not only of the developing world but also of the world itself. Despite the recent debate on the sources of the growth of these countries, the contribution of their national systems of innovation to certain elements of their economic performance, such as their export performance, is relatively less well-known. Given the fact that exports are an important engine of growth and industrial competitiveness, it is important to analyse the contribution of technology to achieving a fast and sustained growth in exports and especially manufactured exports. Recent studies have shown that the location of export production is moving from the industrialised to the developing countries and that exports by the latter have grown rapidly and diversified away from traditional resource- and labour-intensive products to high technology manufactured products (Lall, 1998). However, this development has remained concentrated in a handful of countries. The precise reasons as to how these countries have emerged as exporters of high technology manufactured products and the extent to which their domestic innovative activities have contributed to this performance need to be examined.

This study is divided into six sections. First, I analyse the changing structure of world exports of manufactured products and the place of developing countries in this structure. Second, I develop a consistent and rigorous series of data on the export of high technology products from developing countries. This is accomplished by applying the OECD definition of high technology to the COMTRADE database available on-line from the UN Statistics Division. Employing the data series thus developed, I measure the changing technological complexion of world exports of manufactured products. This section also discusses in detail the various datasets that are available for such a measurement. The third section focuses on three main features of the export of high technology products from developing countries: the concentration of these high tech exports in approximately five countries; the catching up of developing countries; and the competitiveness of high tech exports both at the country level and within each country at the product level. In the fourth section I attempt to verify the "statistical artefact" hypothesis. Three separate indicators are used for verifying the hypothesis. First, I analyse product specialisation in both developed and developing countries. Second, I analyse the patenting record of the developing countries especially with respect to high technology areas. Third, I analyse certain summary indicators of high tech competitiveness. And finally, I analyse the case of the largest exporter of high tech products from the developing world.

It is no longer correct to classify all developing countries as a homogeneous category, especially in terms of their respective technological development. Since the quantum and composition of a country's exports is considered to be one of the best indicators of technological development, I use export performance to classify the developing countries. Before doing so, I present some general features with respect to the trade performance of developing countries. In terms of science and technology (S&T), only 50 countries in the world are active. In 1994, 98 per cent of the world R&D spending, 95 per cent of the total number of scientists and engineers employed worldwide, and 99 per cent of the patents issued (by both the United States and European patent offices) were from these 50 countries.[1] The 50 countries comprise a very heterogeneous group, which includes all the developed countries, most of the newly industrialised countries, and some of the developing countries and countries in transition. These 50 countries, on an average, have an economic growth rate which is much higher than that of the remaining 130 countries: between 1986 and 1994, the growth rate of the 50 countries was 2.4 per cent, or about three times higher than that of the rest of the world. Countries lacking the human capital to make effective use of their natural resources, productive systems, or imported capital goods do less well

14 INNOVATION, LEARNING, AND DYNAMISM

than those which have no natural resources, but which do have educated populations, knowledge, and know-how.

2 Exports of manufactured products from developing countries

Several points need to be mentioned. First, developing countries are increasingly becoming exporters of manufactured products. For instance, manufactured products now account for very nearly three-quarters of the total exports of these countries. Second, the share of developing countries in total world exports of manufactured products has also been increasing (fig. 2.1). Third, the export structure is increasingly moving towards technology-intensive products such as capital goods (table 2.1). Fourth, as such this export performance is not widespread, but is concentrated in a handful of the more developed of the developing countries. In

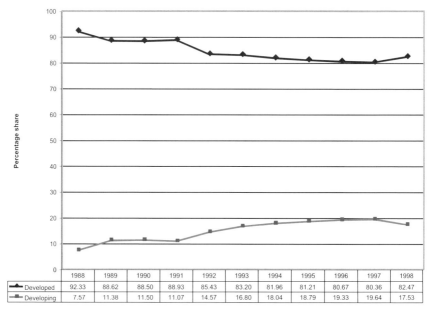

Figure 2.1 Share of developing countries in world exports of manufactured products.
Source: INTECH (2000).

Table 2.1 Structure of exports of developing countries, 1988 and 1998

Standard International Trade Classification (SITC)	Category	1988 (percentage share)	1998 (percentage share)
0	Food and live animals	10	8
1	Beverages and tobacco	0	1
2	Crude materials, inedible, except fuels	7	4
3	Fuel lubricants	6	6
4	Animal and vegetable oils, fats and waxes	2	1
5	Chemicals and related products, n.e.s.	3	6
6	Manufactured goods classified chiefly by material	19	15
7	Machinery and transport equipment	29	39
8	Miscellaneous manufactured articles	23	17
9	Commodities and transactions not classified elsewhere in SITC	0	3
Total		100	100

Source: INTECH (2000).
Note: n.e.s., not elsewhere specified.

other words, most of the developing countries are still not participating in export trade to any great extent. In fact, just 10 countries contribute approximately 90 per cent of the total exports of manufactured products from the developing world and most of them are from the high performing East Asian region (table 2.2).

India has emerged as a leading exporter of computer software products: the rate of growth of software exports has averaged around 50 per cent per annum in the last 10 years or so. However software exports are termed high tech services and, since my study, based essentially on the COMTRADE dataset, is confined to high tech manufactured products, this issue is not analysed here. Nevertheless, the rapid growth of exports of high tech services is slowly emerging as an important area of inquiry. It is against this background that I measure the changing technological complexion of world trade.

Table 2.2 Value of exports of manufactured products[a] from developing countries, 1991–1997

Rank	Country	1991	1992	1993	1994	1995	1996	1997
1	China	NA[b]	66,756,208	73,831,029	99,524,547	124,843,984	127,259,126	155,907,301
2	Korea	66,594,157	70,863,334	76,268,605	89,041,263	114,394,669	114,920,742	118,240,707
3	Singapore	42,716,484	48,581,853	57,950,906	79,504,931	99,037,196	104,120,080	104,753,483
4	Mexico	13,820,892	32,802,591	38,679,112	46,838,911	61,634,515	74,699,958	88,818,438
5	Malaysia	20,791,987	26,232,553	32,809,037	43,312,385	55,084,725	59,284,092	60,179,052
6	Thailand	18,558,400	21,671,504	26,354,921	32,637,019	41,140,927	39,670,035	41,078,884
7	Brazil	17,213,967	20,263,187	22,635,767	23,735,481	24,582,607	25,188,181	27,969,348
8	India	12,816,947	15,135,526	16,291,168	19,978,984	23,184,129	24,090,966	25,702,920
9	Indonesia	11,814,255	16,059,531	19,438,809	20,675,496	22,956,805	25,553,900	22,489,737
10	Philippines	6,169,838	4,039,489	4,664,959	5,739,295	7,051,196	17,005,027	11,134,693
	Total for the above 10	210,396,926	322,405,775	368,924,313	460,988,312	573,910,752	611,792,108	656,274,562
	Total for all developing countries	243,900,252	368,811,528	418,793,235	519,142,360	644,756,480	685,948,116	727,384,110
	Percentage share of the top 10 in the total developing country exports	86.26	87.42	88.09	88.80	89.01	89.19	90.22

Source: INTECH (2000).
Note: Values are in thousands of US$.
a. Manufactured exports are those exports defined in sections 5 through 8 minus division 68 (namely non-ferrous metals) plus group 891 (arms and ammunition).
b. China has been listed in COMTRADE, Revision 3, only since 1992.

3 Changing technological complexion of world trade in manufactured products

It is hypothesised that the technological character of world trade is becoming increasingly complex: table 2.1 shows that much of the world export of manufactured goods consists of products, such as machinery, with a high technology content. The first discussion of this topic was by Kravis and Lipsey (1992) who showed that, for the *market economies*[2] as a whole, trade in manufactured products has been shifting out of low tech goods into high tech products. However, their study only covered the period up to the mid-1980s and was based on UN trade data, while their definition of high technology was based on that of the OECD (1986), though the authors also state that the terms high, medium, and low technology are used throughout their study as shorthand for high, medium, and low R&D intensity.

The first study based on developing countries is that by Lall (1998).[3] The major conclusions of his study are as follows:

"Developing countries are rapidly increasing their shares of manufactured trade, not just in labour-intensive products, but also in capital- and skill-intensive ones; their shares are rising particularly rapidly in the high technology area. However, manufactured exports remain highly concentrated in the developing world, with a few countries dominating all forms of export. Within the successful exporting countries, there are significant differences in the technology content of exports."

However, the terms high, medium, and low technologies are not really defined in any objective sense; only a list of products in each of the three technology categories is given.

In this sense there is a gap in the literature. So in order to measure the technology content of world trade, I first attempt a definition of high technology and apply this definition to one of the most comprehensive sources of data on international trade, namely the COMTRADE[4] dataset, to arrive at a consistent series of data on exports of high technology products from developing countries. In this study I adopt the UN definition of a developing country. In very specific terms the list of such countries contained in the United Nations Industrial Development Organisation (UNIDO) Yearbook (UNIDO, 1999) is used.

3.1 Definition of high technology

Economists have been attempting to measure the technology content of world trade. This is accomplished in terms of the technology embodied in products exported from a country. Admittedly, it is a difficult exercise and no method is foolproof or perfect. The greatest difficulty is in classifying a product according to the technology content embodied in it and several attempts have been made in the past. In the following I undertake a quick review of the methodologies. This is essential, as I will attempt to show that, depending on the definition of high technology employed, it is possible to arrive at significantly different results.

The first systematic effort in this direction was by Davis (1982). He defined high technology manufactures as those manufactured products that have the highest *embodied R&D* spending relative to the value of shipments. To illustrate, the term "embodied R&D" means an aircraft gets credit for the money spent on R&D by the computer industry which supplied the avionics, as well as for the money spent on R&D by the industry supplying the tyres. Davis used input–output techniques to determine how much of the value of R&D embodied in the intermediate products should be included as an indirect addition to the R&D spent directly to produce the final product. Total R&D was thus the sum of both direct and indirect R&D. Arranging the product groups in order from the highest in technology intensity to the lowest, Davis designated the top 10 as high technology. The product with the tenth highest level of technology intensity was 30 per cent higher than that of the eleventh product and more than double the average of all manufactures. The Davis definition was based on the Standard Industrial Classification (SIC) of the US Department of Commerce and as such this definition could not be applied to international data classified according to the Standard International Trade Classification (SITC). Hatter (1985) has overcome this problem by mapping out a concordance between the SIC and the SITC. This concordance was constructed on the basis of SITC Revision 1. The World Bank (1999) have applied this definition to the UN COMTRADE dataset and have arrived at a series of high technology exports at the aggregate level for over 70 developed and developing countries during the period 1962–1997. In the following I test whether my dataset is sufficiently robust by comparing it with that of the World Bank.

However, there are two major limitations with both the definition (Hatter–Davis) and the dataset (World Bank, 1999). First, the definition being applied to a rather aggregate classification such as the SITC Revision 1 is likely to overestimate the extent of high technology exports from countries. Second, one cannot retrospectively apply a definition based on estimates of embodied R&D in the 1980s to say the 1960s, because many of the items that were considered to be "high tech" in the 1960s are unlikely to be considered "high tech" in the 1980s. More on these points later.

The second definition of high technology is by Hatzichronoglou (1997)

Table 2.3 High tech products list by the OECD

Product description	SITC Revision 3 codes
Aerospace	Sum (7922...7925)
	Sum (71441...71491)
Computing and office equipment	75113, sum (75131...75134), sum (7521...7527), 75997
Electronics and telecommunications	76381, 76383, sum (7641...76492), 7722, 77261, 77318, 77625, 77627, 7763, 7764, 7768, 89879
Electrical equipment	Sum (77862...77865), 7787, 77884
Non-electrical equipment	71489, 71499, 71871, 71877, 71878, 72847, 7311, 73131, 73135, 73142, 73144, 73151, 73153, 73161, 73163, 73165, 73312, 73314, 73316, 7359, 73733, 73735
Scientific instruments	Sum (7741...7742), 8711, 8713, 8714, 8719, 87211, Sum (87412...8749), 88111, 88121, 88411, 88419, 89961, 89963, 89966, 89967
Chemical products	Sum (52222...52269), 5251, 5259, 5311, 5312, 57433, 5911, 5912, 5913, 5914
Pharmaceutical	5413, 5415, 5416, 5421, 5422
Armaments	8911, 8912, 8913, 8919

Source: Hatzichronoglou (1997).
Note: List is based on SITC Revision 3.

who prepared a list of high technology products corresponding to the three-digit SITC Revision 3 classification. This list was "the outcome of calculations concerning R&D intensity (R&D expenditure/total sales) covering six countries (the US, Japan, Germany, Italy, Sweden, the Netherlands)". The number of countries covered is of no great importance since national considerations have no bearing on whether a product is classified as high tech or not. The OECD definition (by Hatzichronoglou) appears to be better than the previous attempts because the products are immediately defined as high tech according to their SITC classification, thus obviating the need for any concordance tables. A list of the high tech products according to this definition is given in table 2.3. As mentioned above, such lists are not perfect: in very specific terms the above definition has the following three limitations:[5]

- First, high tech products cannot be selected exclusively by quantitative methods unless a relatively high level of aggregation is adopted. Resorting to expert opinion does make for extremely detailed lists, but the results cannot readily be reproduced in their entirety by other panels of experts.
- Second, if the choice is not based exclusively on quantitative measurements, it is difficult to classify products in increasing or decreasing order.

- Third, the data are not comparable with other industrial data on value added, employment, or gross fixed capital formation published by other agencies such as UNIDO.

3.2 Empirical results

First of all I compare my dataset with that developed by the World Bank (1999). To recapitulate, the World Bank (WB) arrived at estimates by applying the Hatter–Davis definition of high technology to the COM-TRADE database, but the classification used is the older SITC Revision 1.[6] The main problem, as discussed earlier, is that an aggregate definition is adopted. The WB estimates are therefore likely to overestimate the quantum of high tech exports from all the countries of the world. This can be checked by comparing the WB estimates with INTECH estimates (fig. 2.2).

In figure 2.2, I present the ratio of WB estimates to INTECH estimates for five developing and four developed countries.[7] The ratio is greater than unity for all countries in my sample, thus indicating that the WB figures are clearly overestimates. Second, there are also what I may call "pure and simple" problems with the data. This can be demonstrated by drawing up a list of the top high tech exporters in the world according to

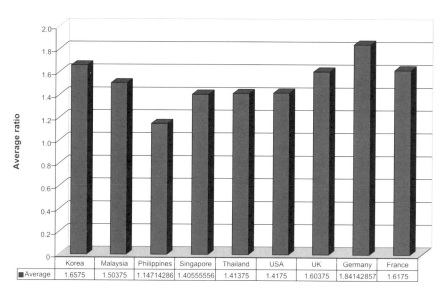

Figure 2.2 Ratio of World Bank estimates to INTECH estimates.
Source: INTECH (2000).

Table 2.4 The leading high tech exporters in the world, 1997

Categories	According to WDI (1999)	According to INTECH (2000)
High exporters (>50%)	Singapore (71%) Malaysia (67%) Jamaica (67%) Ireland (62%) Philippines (56%) Senegal (55%)	Singapore (57%) Malta (56%) Philippines (53%) Malaysia (49%) Ireland (47%) USA (32%)
Medium exporters (25–50%)	Netherlands (44%) USA (44%) Thailand (43%) UK (41%) Australia (39%) Korea (39%) Hungary (39%) Japan (38%) Nicaragua (38%) Sweden (34%) Israel (33%) Mexico (33%) Hong Kong (29%) Switzerland (28%) Denmark (27%) Saudi Arabia (27%) Morocco (27%) Trinidad and Tobago (27%) Germany (26%) Jordan (26%)	Thailand (31%) Korea (27%) Netherlands (27%) Japan (26%)

Source: INTECH (2000).
Note: WDI, World Development Indicators. Figures in brackets indicate the percentage share of high tech exports in total manufactured exports.

each dataset. Once again the WB dataset suggests some unlikely countries (table 2.4).

It is clear that the WB estimates of high tech exports with respect to six countries, namely Jamaica, Senegal, Jordan, Nicaragua, Morocco, and Trinidad and Tobago, are extremely doubtful. In contrast, Malta is the only surprising country on our list.[8] This aspect is further examined by comparing the absolute values of exports for each of these countries according to both datasets (table 2.5).

The foregoing analysis clearly demonstrates that, despite imperfections in the definition, etc., the INTECH dataset is quite suited to measuring the technological content of world trade.

First, world trade is increasingly becoming a trade in high tech prod-

Table 2.5 Comparison of World Bank and INTECH datasets

Country	WDI		INTECH	
	Value of high tech exports (millions of US$)	Share of high tech exports (%)	Value of high tech exports (millions of US$)	Share of high tech exports (%)
Jamaica	619	67	0.509	17.00
Senegal	145	55	Nil	Nil
Jordan	183	26	Nil	Nil
Nicaragua	61	38	Nil	Nil
Morocco	622	27	9.51	0.65
Trinidad and Tobago	296	27	11.43	1.03

Source: INTECH (2000).
Note: Data are based on the values for 1997.

ucts (fig. 2.3) with approximately 22 per cent of the exports of manufactures being of this type. What is more striking is the significant increase in the technology content of exports from developing countries: nearly 27 per cent of the exports from developing countries are now high tech products. In this sense, my finding is very similar to that of Lall (1998). Second, the share of developing countries in high tech exports has shown a dramatic increase: it has increased from about 8 per cent in 1988 to about 23 per cent in 1997 (fig. 2.4).

Third, the exports of developing countries (1988–1998) have grown at a much faster rate of 31 per cent per annum compared with the 15 per cent per annum of the developed countries. This higher rate of growth is not only at the aggregate level but also across the various products. In fact, except for pharmaceuticals, the growth performance of the developing countries is much higher.

The World Bank (2000) has (in its latest *World Development Indicators 2000*) changed its 1999 methodology to that which I have employed in this study (OECD) and there are therefore no differences now between them.[9]

4 Main features of high technology exports

I consider four main features:
- The concentration of high tech exports in a few developing countries
- The catching up of developing countries in high tech exports

EXPORTS OF HIGH TECHNOLOGY PRODUCTS 23

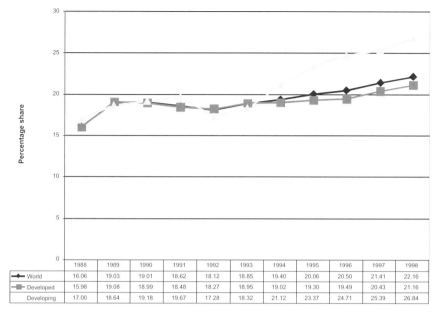

Figure 2.3 Trends in high technology content of world exports of manufactured products.
Source: INTECH (2000).

- The competitiveness of high tech exports from developing countries
- Net export ratio

4.1 Concentration of high tech exports

There is considerable concentration of high tech exports within both the developed and developing countries. Approximately three-quarters of the total exports of high tech products in the world are produced by just 10 countries, of which five are developing. Two rankings are possible, one in terms of the absolute value of exports and the other in terms of the relative share of high tech exports in the exports of manufactures from each country (both developed and developing). The second ranking refers more to the high tech export intensity. If one uses the first indicator, the situation shown in table 2.6 emerges.

However, if one uses the second indicator, namely the relative share of high tech exports, and ranks all the countries according to their high tech export intensity, then the arrangement is entirely different (table 2.7).

It may well be that some of these countries, especially those from the

24 INNOVATION, LEARNING, AND DYNAMISM

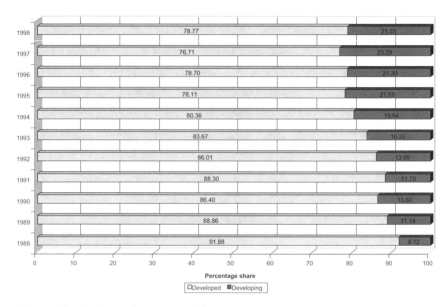

Figure 2.4 The increasing share of developing countries in total world exports of high tech exports, 1988–1998.
Source: INTECH (2000).

developing world and perhaps Ireland from the developed world, are not *real* manufacturers of high tech products. This point will be examined in some detail in sections.

4.2 The catching up of developing countries

At the outset it must be made very clear that I use the term "catching up"[10] in a very different way from the normal sense of the term as used in the literature on technology transfer and development. This is because, although one country may have caught up with another country in terms of export share, its ability to maintain that share over a long period of time is questionable as market shares are inherently unstable. Second, the catch-up is defined purely in terms of the quantum of gross exports and not net exports. The issue of net exports is treated separately in section 4.4. With these caveats, an examination of the data shows the following:
- The developing countries have been catching up with the developed countries. This can be shown by taking a ratio of developing country exports to developed country exports (fig. 2.5). Within approximately a decade the share of developing country exports has increased by as much as 20 percentage points.

Table 2.6 Concentration in exports of high tech products

Year	Share of the top 10[a] high tech exporting countries (in per cent)	Concentration within each category (in per cent)	
		Share of top five developed countries	Share of top five developing countries
1988	68	66	87
1989	77	76	87
1990	75	74	86
1991	78	78	82
1992	78	76	85
1993	77	76	86
1994	77	75	86
1995	76	73	87
1996	74	72	82
1997	75	72	85
1998	77	72	95

Source: INTECH (2000).
Note: Data are based on absolute values.
a. The top five developed countries are the USA, Japan, Germany, the UK, and France and the top five developing countries are Singapore, Korea, Malaysia, China, and Mexico.

- As a corollary of the above, the catching up is concentrated in certain specific product groups, which are of course the most important within the export basket of these countries (fig. 2.6).
- The maximum catch-up has occurred in the computing and office equipment category followed by electronics. In contrast, the least catch-

Table 2.7 Ranking of developed and developing countries according to high tech export intensity

Developed countries			Developing countries		
Rank	Name	Average export intensity[a]	Rank	Name	Average export intensity[a]
1	Ireland	42.82	1[b]	Singapore	49.78
2	USA	31.90	2	Malaysia	43.13
3	Japan	25.11	3	Philippines	38.33
4	UK	24.96	4	Thailand	22.79
5	Netherlands	21.71	5	Korea	21.62

Source: INTECH (2000).
a. Average during the period 1988–1998.
b. Malta, which has an average export intensity of 55.72 per cent, is the highest in the world. For the time being, this has been excluded (see endnote 8).

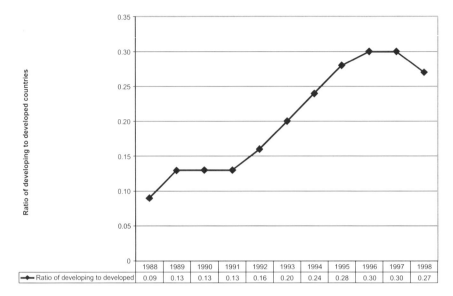

Figure 2.5 The catching up of developing countries with respect to high tech exports (at the aggregate level).
Source: INTECH (2000).

up has occurred in the non-electrical machinery and pharmaceutical products categories.
- Specific developing countries have caught up with specific developed countries. The United States is the largest exporter of high tech products in the world. The top developing and developed countries (in terms of high tech export intensity) are compared with the United States in fig. 2.7. It is seen that the maximum catch-up is by Singapore. As a matter of fact Singapore exports (as regards the absolute level of exports) more high tech products than the United Kingdom, Germany, or France.
- Specific developing countries have caught up with specific developing countries. Malaysia is the developing country that has caught up most (fig. 2.7): it has almost caught up with Korea and is fast catching up with Singapore.

4.3 *Competitiveness of high tech exports*

Export success is generally regarded as a measure of the competitiveness of a country's industries. It has been noted that markets for high tech products are growing faster than those for other products because of the

EXPORTS OF HIGH TECHNOLOGY PRODUCTS 27

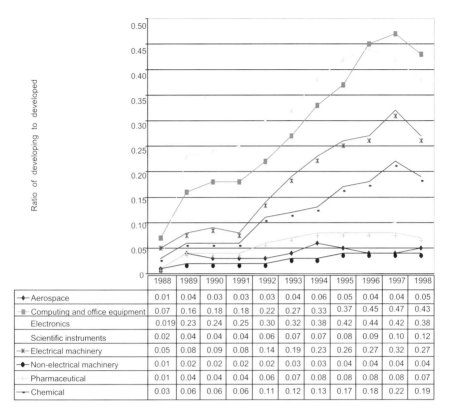

	1988	1989	1990	1991	1992	1993	1994	1995	1996	1997	1998
♦ Aerospace	0.01	0.04	0.03	0.03	0.03	0.04	0.06	0.05	0.04	0.04	0.05
■ Computing and office equipment	0.07	0.16	0.18	0.18	0.22	0.27	0.33	0.37	0.45	0.47	0.43
Electronics	0.019	0.23	0.24	0.25	0.30	0.32	0.38	0.42	0.44	0.42	0.38
Scientific instruments	0.02	0.04	0.04	0.04	0.06	0.07	0.07	0.08	0.09	0.10	0.12
✳ Electrical machinery	0.05	0.08	0.09	0.08	0.14	0.19	0.23	0.26	0.27	0.32	0.27
● Non-electrical machinery	0.01	0.02	0.02	0.02	0.02	0.03	0.03	0.04	0.04	0.04	0.04
+ Pharmaceutical	0.01	0.04	0.04	0.04	0.06	0.07	0.08	0.08	0.08	0.08	0.07
— Chemical	0.03	0.06	0.06	0.06	0.11	0.12	0.13	0.17	0.18	0.22	0.19

Figure 2.6 The catching up of developing countries by product level.
Source: INTECH (2000).

higher income elasticity of demand and greater scope for product innovation and productivity increases (Lall, 1998; UNCTAD, 1999). The implication of this is that it is easier for a country to sustain export growth if it can establish a competitive position in these products than in the slow growing non-high tech products. So in the following I analyse the competitiveness of high tech products from developing countries. This is measured by computing indices of revealed comparative advantage (RCA) (fig. 2.8).[11]

The RCA indices are computed not only at the aggregate level but also for each of the subsectors within. The developing countries are more competitive than the developed countries for most years and especially since the mid-1990s.

Singapore and Malaysia have the highest RCAs of the developing countries (table 2.8).

28 INNOVATION, LEARNING, AND DYNAMISM

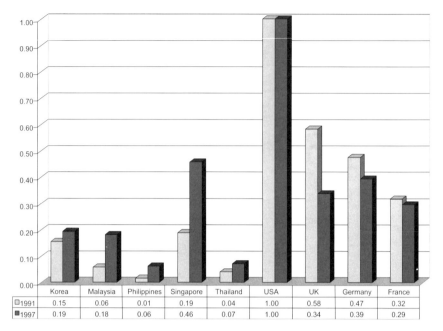

Figure 2.7 The catching up of specific countries with the United States.
Source: INTECH (2000).

In the above, I have presented the main characteristics of high tech exports from developing countries. In section 5, I will attempt to analyse whether this performance of the developing countries is real or is a mere statistical artefact.

4.4 Net export ratio

The entire analysis thus far has been with respect to gross exports. However, it is important to view the position of developing countries as regards net exports, so I define a ratio of net exports (high tech exports minus high tech imports) to gross exports. Being a ratio, it facilitates inter-spatial comparisons. The ratio, though negative as far as the developing countries are concerned, has been improving over time (fig. 2.9). In contrast, the ratio for the developed countries has been worsening.

Further, all the countries are ranked according to the average net export ratio during the period under consideration. Singapore, Korea, and Malaysia have positive net export ratios and figure in the top 19 countries. Thailand is also in the top 19, but has a negative net export ratio. Only the Philippines does not figure in the list (fig. 2.10).

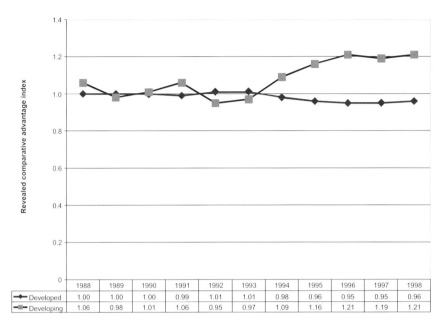

Figure 2.8 Competitiveness of developed and developing countries in high tech exports.
Source: INTECH (2000).

Table 2.8 RCA indices of leading high tech exporters from the developing world

Year	China	Mexico	Korea	Philippines	Thailand	Malaysia	Singapore
1988	–	–	0.99	–	1.02	2.53	–
1989	–	0.53	0.94	–	1.00	2.02	1.96
1990	–	0.44	0.95	–	1.10	2.01	2.10
1991	–	0.46	1.04	1.74	1.12	2.05	2.15
1992	0.36	0.62	1.09	1.52	1.22	2.15	2.47
1993	0.38	0.62	1.08	1.61	1.10	2.18	2.46
1994	0.43	0.72	1.17	1.63	1.22	2.28	2.61
1995	0.52	0.76	1.30	1.75	1.22	2.30	2.70
1996	0.61	0.77	1.17	2.85	1.42	2.17	2.72
1997	0.61	0.82	1.24	2.47	1.44	2.29	2.66
1998	0.69	0.87	1.22	–	–	2.48	2.66

Source: INTECH (2000).

30 INNOVATION, LEARNING, AND DYNAMISM

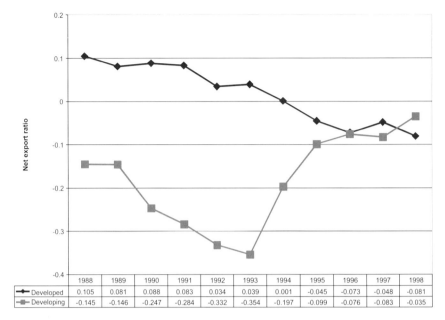

Figure 2.9 Net export ratio of developed and developing countries, 1988–1998.
Source: INTECH (2000).

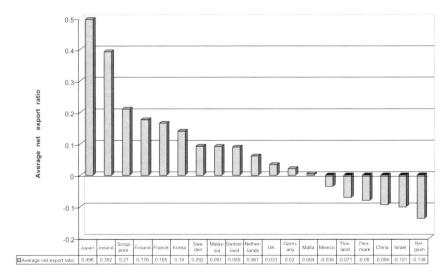

Figure 2.10 The top countries in terms of average net export ratio, 1968–1998.
Source: INTECH (2000).

5 Verification of the statistical artefact hypothesis

It has been observed by Lall, Albaladejo and Aldaz (1999) that the same product can use very different processes in different locations, exploiting different sources of competitive advantage. For instance, according to them, "semiconductor exports may involve complex design and fabrication in one country (and so be genuinely high technology) and only assembly and packaging in another (and so be low technology). Many high tech exports from developing countries are based on the relatively simple, labour-intensive assembly of imported components." In other words, the high tech exports from developing countries are a mere statistical artefact and are not really real. I propose to verify this hypothesis by examining:

- The *product specialisation* within each of the eight high tech product groups by both developed and developing countries. In very specific terms, I will examine whether the products specialised in by the developing countries are very different from those of the developed countries. This will be done at a very high level of disaggregation. If after the analysis I reach the conclusion that the nature and kinds of products manufactured by developing countries and the developed countries are not very different, it will be extremely difficult to maintain the statistical artefact hypothesis. Needless to add, verifying by this method has some limitations. Of course we can verify it if developing countries are specialised in the same high tech products, but even if these products have the same name they are not necessarily similar and they could have a different technological content. For instance, a microphone or a television camera can have a high range or a low range. The trade classification does not make any distinction, but the technological knowledge and know-how necessary to produce these products are quite different.
- The *innovative activity* of the developing countries with respect to high technology areas. If a country has real capability to design and manufacture (rather than assemble) high technology products, then that capability should be reflected in terms of its patenting record.
- The *indicators of high tech competitiveness* developed by the Technology Policy Assessment Center (TPAC).[12] TPAC researchers define four indicators of high tech competitiveness: (i) national orientation, which indicates a country's commitment to technology-based development along the following dimensions: government policy, political stability, entrepreneurial spirit, and acceptance of the idea that development should be technology-based; (ii) socio-economic infrastructure, which indicates the strength of each nation's educational system, mobility of capital, and encouragement of foreign investment; (iii) technological infrastructure, which captures the strength and contributions

32 INNOVATION, LEARNING, AND DYNAMISM

Table 2.9 Structure of high tech exports of developing and developed countries (%)

	World	Developing	Developed
Electronics and telecommunications	34.66	47.18	31.32
Office and computing	27.33	31.57	29.69
Aerospace	13.05	3.10	15.00
Scientific instruments	10.15	11.41	3.64
Electrical machinery	2.63	2.23	2.68
Non-electrical machinery	3.75	0.57	4.37
Pharmaceuticals	3.87	1.27	4.40
Chemical	4.56	2.67	8.88

Source: INTECH (2000).

of a nation's scientific and engineering manpower, its electronic data processing purchases, the relationship of its R&D to industrial application, and its ability to make effective use of technical knowledge; and (iv) productive capacity, which is the capability to manufacture technology-intensive products. It combines the value of electronics production with three survey items related to manufacturing and managerial capabilities to measure the amount and efficiency of resources available. An examination of the scores obtained by the developing countries enables one to draw some informed conclusions about their technological capability to manufacture and export high tech products.

5.1 Product specialisation in developed and developing countries

At the aggregate level, developing countries specialise in the export of electronics items and office and computing equipment (table 2.9).

Developed countries, in contrast, have a much better distributed export structure. In addition to the above two items, they also export sizeable quantities of both aerospace and scientific instruments. So, the impression is that the export structure of developing countries concentrates on two items. I will now examine whether the export structures of the two groups are very different within these two groups of electronics and office and computing equipment (table 2.10).

It is seen that both the developed and developing countries are increasingly specialising in the manufacture and export of components and both electronic and office equipment parts. This is especially so in the case of office and computing equipment where very nearly one-third of the exports of both groups are automatic data processing machine components. Based on this, it can be inferred that the degree of specialisation of both developing and developed countries is similar. However, it is in-

Table 2.10 Specialisation of developed and developing countries within electronics and office and computing equipment, 1991 and 1997

Electronics and telecommunications	Developing 1991	Developing 1997	Developed 1991	Developed 1997	Office and computing equipment	Developing 1991	Developing 1997	Developed 1991	Developed 1997
Video recording or reproducing apparatus	10.20	5.62	6.58	2.79	Auto typewriter, word processing machine	0.11	0.07	0.34	0.07
Other sound reproducing apparatus	0.61	1.93	2.50	1.07	Electrostatic photocopying (direct process)	0.05	0.14	0.17	0.09
Line telephones, etc.	4.97	6.38	9.27	10.36	Electrostatic photocopier (indirect process)	0.21	1.64	4.52	3.56
Microphones, loudspeakers, amplifiers	3.18	3.30	3.32	1.97	Non-electrostatic photocopying apparatus of the contact type	0.00	0.01	0.15	0.03
Transmission apparatus for radio telephony, etc.	3.38	3.82	7.07	11.36	Analogue or hybrid computers	0.04	0.14	0.15	0.03
Telecommunications equipment, n.e.s.	2.20	2.91	10.03	4.04	Digital computers	14.20	5.45	0.56	1.17
Parts of line telephony	1.40	1.62	7.18	8.16	Digital processing units	2.37	5.02	7.02	7.43
Parts of microphone, etc.	0.37	0.40	0.37	0.27	Input or output units	43.59	21.24	18.91	19.32
Printed circuits	2.55	4.59	5.85	3.48	Storage units	4.60	31.11	7.16	14.85
Panels, etc., for voltage not exceeding 1000 V	0.46	0.80	3.99	2.87	Parts of automatic data processing machines	34.83	35.20	39.42	38.19
Optical fibre cables	0.02	0.17	0.57	0.73	Total	100.00	100.00	100.00	100.00
Microwave tubes	0.03	0.06	0.37	0.42					
Other valves and tubes	0.01	0.12	0.39	0.43					
Diodes, transistors, etc.	9.13	7.73	4.74	5.27					
Electronic integrated circuits and microassemblies	55.49	55.68	29.33	37.57					
Peizo-electric crystals	5.72	3.88	2.64	3.18					
Recorded media n.e.s.	0.29	0.98	5.81	6.04					
Total	100.00	100.00	100.00	100.00					

Source: INTECH (2000).
Note: The percentage share of exports within each of the two categories is shown.

correct to state that, while developing countries specialise in the export of finished products, developing countries specialise in the export of just-assembled finished products. In fact, we find that both are exporters of components or parts, at least in these two categories.

A third way of checking the degree of specialisation of both groups is to rank the top 10 products in terms of the value of exports. This is done for both developing and developed countries for the two years 1991 and 1997, but on the basis of the ranks obtained in 1997 (table 2.11).

It is seen that the top two items in both groups are exactly the same. The top 10 items account for over three-quarters of developing country exports, while they account for only about a half of developed country exports. In short, developing country exports are concentrated in a few items. Once again, many of the developing country exports are parts and components.

Thus my analysis of product specialisation clearly shows that both developing and developed countries are exporting, by and large, the same types of products. The data do not support the conclusion that developing countries are specialising in the export of technologically less sophisticated products (within the high tech sector).

5.2 Patenting record of developing countries

Patent data are a rather good, if imperfect, indicator of the technological capability of a country. Pavitt and Soete (1980) have used patent data to analyse the relative "competitiveness" of various countries, to construct "revealed technology advantage" indexes for various countries, and to describe and contrast the international location of inventive activity in different industries.

Patents are not the only way to protect intellectual property rights: copyrights and trade secret laws also protect certain types of intellectual property. For example, computer programs and integrated circuit configurations are usually protected by copyright. However, because patenting is the primary form of intellectual property protection, patent data are considered to be the most available, objective, and quantitative measure of innovative output (Griliches, 1990). Thus a country's patenting activity is an indicator of the strength of its research enterprise and technological strengths, both overall and in particular fields of technology.

I start first with overall patenting activity by examining the trends in the number of patents granted in the United States to researchers from different countries since 1985 (fig. 2.11). Since the patent counts for the countries in this study cover such a wide range, I have plotted them in the logarithmic scale.

The patenting rate for many of the developing countries has increased

Table 2.11 The top 10 high tech exports by developing and developed countries

	Developing countries			Developed countries	
	1991	1997		1991	1997
Electronic integrated circuits	13,914,501	49,127,995	Electronic integrated circuits	28,987,309	79,558,903
Parts of automatic data processing machines	5,804,825	25,343,488	Parts of automatic data processing machines	35,770,453	58,640,853
Storage units	766,647	22,393,643	Aircraft, etc. ULW > 500 kg	41,340,065	37,889,184
Input or output units	7,265,324	15,289,681	Digital processing and storage units	17,164,319	29,668,113
Diodes, transistors, etc.	2,288,920	6,822,845			
Line telephones, etc.	1,245,314	5,626,230	TV, radio transmitters	6,982,483	24,066,462
Video recording and reproduction apparatus	2,556,952	4,962,619	Storage units	10,654,333	23,454,389
			Input or output units	15,576,291	22,801,798
Printed circuits	639,905	4,050,710	Line telephone, etc.. equipment	9,159,223	21,934,176
Digital computers	2,366,679	3,920,626			
Digital processing and storage units	394,448	3,610,907	Parts of line telephony equipment	7,091,715	17,278,829
Total for the above 10 items	37,243,516	141,148,743	Parts of jet, turbo-prop engines	9,273,437	13,172,583
			Total for the above 10 items	181,999,628	328,465,291
Total high tech exports	47,961,460 (78)	184,746,341 (76)	Total high tech exports	362,052,787 (50)	608,304,139 (54)

Source: INTECH (2000).
Notes: Figures in parentheses indicate the percentage share of the top 10 items in the overall high tech exports of each category. Values are in thousands of US$. ULW, ultra light weight.

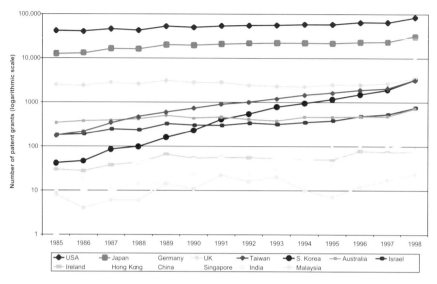

Figure 2.11 US patent activity by inventor country and grant year, 1985–1998. *Source*: USPTO (1999).

dramatically in recent years. This is particularly true for Korea and Taiwan.[13] If the current patent growth rates hold, within a few years both Taiwan and Korea will surpass the United Kingdom in the number of United States patents granted annually. China and Singapore show the fastest growth in United States patents, though the number of patents granted to each is still small. Malaysia has the least number of patent grants, but this is showing an increasing trend since the mid-1990s.

A recent study by the Office of Technology Policy (1998) has examined the global patenting trends in five high technology sectors: health, advanced materials, the automotive industry, information technology, and express package transportation logistics. Though these sectors do not coincide with our definition of high technology, there are two common sectors, the pharmaceutical industry and information technology. In order to analyse the patenting records of the various countries in each of the five sectors, the study considers three five-year periods: 1982–1986, 1987–1991, and 1992–1996.

- The study shows that the technological capabilities of Korea and Taiwan are increasing, with their growth strength most evident in advanced materials and information technology. Since 1995, both Korea and Taiwan have exceeded the United Kingdom and Germany in information technology patents issued.

- On the other hand, Malaysia and Singapore do not have enough patenting activity in any of these sectors to be identified as high tech. This fact could reflect these countries' lack of a significant indigenous R&D capability.
- This means that the exports of high technology products from these countries are largely contributed by multinational corporations (MNCs) with very little local R&D effort. However, this situation is changing rapidly: Singapore in particular is investing very heavily in domestic R&D and has also taken a number of steps to strengthen its domestic innovative activity.

These findings are supplemented by my own analysis of the patenting records of the top 15 most common United States patent classes[14] for inventors from Korea, Singapore, and Malaysia for the period 1994–1998. Similar data for other developing countries such as Thailand and the Philippines are not immediately available (table 2.12).

It is clear from table 2.12 that, while in the case of both Korea and Singapore many patents are in the area of high technology (loosely defined), the same cannot be said about Malaysia. Thus from my analysis of the patenting records of developing countries, it appears that Korea and increasingly Singapore have design capabilities in many areas of high technology, especially information technology. Malaysia, Thailand, and the Philippines, in contrast, do not appear to have such capabilities, though Malaysia does seem to have a better record in this area than the other two countries.

It is very instructive to examine the relative contributions of both foreign and domestic firms to the total number of patents issued.[15] My analysis (based on the United States Patents and Trademarks Office, USPTO) shows that local firms account for all the patents procured by Korean, Taiwanese, and Chinese inventors. In the case of Singapore, 58 per cent of the patents are issued to local firms. Of the patents taken out by Malaysian inventors, 75 per cent are individually owned, while the remaining 25 per cent belong to one foreign company. Thailand and the Philippines have an insignificant number of patents.

5.3 Indicators of high tech competitiveness

As mentioned earlier there are four such indicators. The scores on these indicators for the leading developing countries and two of the developed countries, namely the United States and Japan, are presented in table 2.13. The following inferences can be drawn from this table.
- National orientation: This indicator attempts to identify those nations whose business, government, and cultural orientation encourage high technology development. Singapore has the highest score on this indi-

Table 2.12 Top 15 most common US patent classes for inventors from Korea, Singapore, and Malaysia

Korea	Cumulative number of patents 1994–1998	Singapore	Cumulative number of patents 1994–1998	Malaysia	Cumulative number of patents 1994–1998
Semiconductor device manufacturing process	741	Cumulative total for all patent classes	8,747	Cumulative total for all patent classes	406
Television	584	Semiconductor device manufacturing process	58	Communications: electrical	5
Static information storage and retrieval	468	Active solid state devices	23	Horizontally supported planar surfaces	4
Electrical connectors	328	Electrical connectors	19	Optical wave guides	4
Dynamic information storage retrieval	272	Television	14	Games using tangible projectiles	4
Dynamic magnetic information storage or retrieval	271	Electric lamp and discharge devices	11	Printing	3
Refrigeration	176	Electricity: measuring and testing	10	Conveyors: power-driven	3
Television signal processing	168	Information processing systems	10	Information processing systems	3
Optics: systems and elements	153	Coded data generation or conversion	9	Hydraulics and earth engineering	3
Electric lamp discharge devices	150	Electricity: electrical systems and devices	9	Wells	2
Radiation imagery chemistry	147	Typewriting machines	9	Dispensing	2
Electric heating	145	Wells	8	Food or edible material: processes, compositions	2
Miscellaneous active electrical nonlinear devices, circuits and systems	132	Electric power conversion systems	8	Stock material or miscellaneous	2
		Chemistry: molecular	8	Synthetic resins	2
		Incremental printing of symbiotic information	7	Baths, sinks, etc.	2
Information processing system organization	132	Electrical computers and data processing systems	6	Brushing, scrubbing	1
Liquid crystal cells, elements and systems	119			Cumulative total for the above 14 categories	1
Cumulative total for the above 14 categories	3,986	Cumulative total for the above 14 categories	209	Cumulative total for all patent classes	41
					69

Source: USPTO (1999).

Table 2.13 Indicators of technological competitiveness

	National orientation	Socio-economic infrastructure	Technological infrastructure	Productive capacity
Hong Kong	74.4	69.6	23.0	43.0
Singapore	92.7	73.3	40.5	54.6
South Korea	81.9	69.6	42.6	46.4
Taiwan	81.1	74.5	37.4	43.1
China	62.3	46.4	38.6	33.2
India	52.4	46.4	33.0	38.6
Indonesia	62.5	49.5	25.3	24.8
Malaysia	81.1	63.7	34.3	47.5
Japan	86.3	72.7	83.7	92.7
USA	69.9	84.0	87.5	89.8

Source: National Science Foundation (1995).

cator. The country has a clear-cut government policy on the development of high technology industry. This point will be further discussed later.
- Socio-economic infrastructure: This indicator assesses the underlying physical, financial, and human resources needed to support modern technology-based economies. Singapore once again has a high score reflecting its small size and the national plan for technology-based growth.
- Technological infrastructure: This indicator evaluates the institutions and resources that contribute to a nation's capacity to develop, produce, and market new technology. Among the developing countries, Korea has the highest score. Singapore's technological infrastructure was rated nearly as high as that of Korea and better than that of Taiwan.
- Productive capacity: This indicator evaluates the strength of a nation's current, in-place manufacturing infrastructure as a baseline for assessing its capacity for future growth in high tech activities. Singapore once again has the highest score among the developing countries.

However, the scores secured by the developing countries on all the indicators except "national orientation" are much lower than those obtained by the developed countries.

In sum, my analysis shows that, as regards product specialisation, both developing and developed countries specialise in the export of parts and components. However, the high tech exports of developing countries are concentrated in 10 products, while those of the developed countries are more spread out. With respect to patenting, although only two of the developing countries have a significant number of patents in seemingly high tech areas, the number of patents procured by these select countries shows a definite increase. However, some of the developing countries, which are major exporters of high tech products, have no patents at all. This means that they are merely acting as subcontractors to large MNCs and as such they do not possess any real technological capability. Finally, my analysis of certain summary measures of high tech competitiveness once again confirms the view that only two of the top five developing countries possess real capability in the design and manufacture of high tech products. To confirm that even these countries are indeed real exporters of high tech products, we require the following two sets of data: (i) the share of local content in the exports; and (ii) the share of local firms (as against subsidiaries of MNCs) in the total exports of high tech products. However, these figures are hard to obtain.

6 Case study of a leading exporter from the developing world

Singapore is the leading exporter of high tech products not only in the developing world but also in the world itself. The general impression or feeling is that many of the goods exported from Singapore come from the subsidiaries of MNCs that operate in that country. An analysis of Singapore's growth process over the last three decades or so reveals at least six characteristics (Huff, 1999), namely sustained rapid growth, strong export orientation, high savings and investment, low inflation, low government consumption, and fundamental structural transformation. This situation has played an important role in enhancing the technological capability of this small, but open, island nation. This role has been achieved through two separate institutional mechanisms: first, through the establishment of a specialised institution supporting technical change, namely the National Science and Technology Board (NSTB); and second, through an imaginative use of industrial policies to increase spillover from the numerous foreign manufacturing companies. These mechanisms are analysed in some depth below.

6.1 *Role of innovation policies*

In order to further develop Singapore's S&T capabilities and support industrial clusters, the second five-year National Science and Technology Plan 2000 was introduced in 1996. The NSTB is central to the plan. Established in 1991, the Board is responsible for developing Singapore's S&T capabilities through strengthening its technology infrastructure and supporting selected industrial clusters. The following clusters have been selected for further development:

- Chemicals and environmental technology
- Electronics components and systems
- Life sciences (biotechnology, food, and agro technology)
- Manufacturing and engineering systems

These industries are supported by 13 government research institutes and centres which form a bridge between basic research carried out in the universities and applied R&D undertaken by the industry.

The NSTB has two main goals. The first goal is to promote and develop a techno-entrepreneurial environment in the country, which will lead to vibrant technology-oriented companies. The second goal is to develop capability through promoting and strengthening the technology infrastructure in the country. The NSTB has established a network of research institutes and centres (in 1998 they collaborated in approximately 310 joint projects with the industry, filed 67 patents, concluded 28 licensing agreements, and collaborated in a combined total of four high tech start-ups) and also works closely with tertiary institutions to ensure that collectively the industry can rely on a firm knowledge base to support current and future needs for manpower and technology. It has also put in place the following measures and programmes to seed and grow technology enterprises in Singapore: (i) the Business Angel Fund; (ii) the Technology Incubator Programme; (iii) the Technology Development Fund; (iv) the Innovator Assistance Scheme; (v) the Incubation Support scheme; (vi) the Techno-entrepreneur Home Office Scheme; and (vii) the Techno-entrepreneur Investment Incentive Scheme.

Consequent to these efforts, both gross expenditure on R&D (GERD) and its intensity has shown some dramatic increases in the last few years (fig. 2.12). Singapore's research intensity, though lower by the international standards of the developed countries, is fast catching up.

Of the total R&D investment, nearly two-thirds is contributed by the private sector mainly out of its own funds: the Singapore government only contributes about 10 per cent.

Within the private sector much of the R&D investment is accounted for by the manufacturing sector and within that the electronics industry. In other words, the electronics industry, which is the largest exporter, is also the largest investor in R&D.

Another important point is that local enterprises invest much more than foreign enterprises in R&D in most sectors, though in the manufacturing sector as a whole and within the electronics industry in particular, they account for only half of that expended by foreign enterprises.

The government has also actively promoted various types of technical education and thereby increased the supply of scientists and engineers. For instance, the number of scientists and engineers engaged in R&D has increased from 18.4 (per 10,000 labour force) in 1984 to 65.51 in 1998.

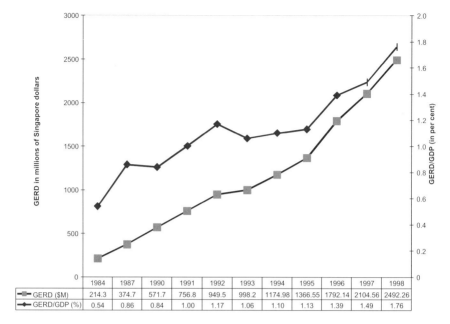

Figure 2.12 Trends in R&D expenditure and in research intensity in Singapore.
Source: National Science and Technology Board (1998).

6.2 Role of industrial policies

Lall (1997) shows that the government in Singapore has actively intervened through various types of industrial policies. The main policy thrusts, according to him, can be divided into those that:
- Deepen the industrial structure: there is a very strong push into specialised high tech industry for export markets, but without protection.
- Raise local content: under its Local Industries Upgrading Programme, the government encourages MNCs to "adopt" a group of small- and medium-sized enterprises (SMEs) and transfer technology and skills to them. It pays the salary of a full-time procurement expert to work for specified periods with the adopted firms and help them upgrade their production and management capabilities to the standards required.
- Support foreign direct investment (FDI) policies: aggressive targeting and screening of MNCs is practised, with direction into high value-added activities. The decisions of MNCs concerning what new technologies to bring into Singapore are influenced by the incentive system, the provision of excellent infrastructure, and the direction offered by the Government of Singapore.

- Raise technological effort: MNCs are targeted to increase R&D.
- Change industrial structure: some public sector enterprises enter targeted areas.

As a result of the combined effect of both technology and industrial policies, there has been a dramatic increase in the share local companies hold in the total industrial investment in the country.

Scientists from the World Technology Evaluation Center (1997) have studied the technological capability of Singapore's electronics industry. According to this study, Singapore is moving more and more towards increased research, with high-volume, low-cost manufacturers (table 2.14). Singapore is moving rapidly into design and development and is increasing investment in research in all areas of electronics. The production of integrated circuits (ICs), IC packaging, printed circuit board (PCB) fabrication, and PCB assembly are well established. The Singapore printed circuit board assembly industry is on a par with (or slightly behind) that of the United States in terms of technology. The industry has increasing design capabilities in all of these areas and is moving into basic materials research to support next-generation component development.

7 Summing up

The purpose of this chapter has been to develop a consistent series of data on the exports of high tech products from developing countries. My analysis has shown that, while the share of developing countries in high tech exports is very large, it is highly concentrated in approximately five countries. In other words, the majority of developing countries take no part in the manufacture or export of high tech products. There is much variation in the technological capability of even the high tech exporting countries. At one end of the spectrum are Korea and (increasingly) Singapore, which have the local capability to design, manufacture, and export high tech items. Malaysia is somewhere in the middle, while Thailand and the Philippines appear to be at the other end with very low (relatively speaking) capability. However, the developing countries as a whole are fast catching up with the developed countries, so it may be unwise to dismiss this performance as a mere statistical artefact.

The study has thus identified some areas for further research: measurement of the import content of these exports and measurement of the relative contribution of MNCs vis-à-vis local companies, because it is well known that in some countries (for instance Malaysia) MNC subsidiaries contribute many of the exported goods.

Table 2.14 Model of Singaporean electronics technology development

Technology stages	Product technologies			
	Integrated circuit	IC package	PCB assembly	PCB
Research	Cell design	Materials	Materials	Materials, plating, wet chemistry
Design	Logic design, module layout	Mechanical characterization, package design	Circuit design, layout	Layout
Deployment	Production engineering	Production engineering	Production engineering	Production engineering

Source: World Technology Evaluation Center (1997).

Notes

This paper is based on a UNU-INTECH research project, Project 99431-1. Thanks are due to Noi Kwanjai for excellent research assistance. I am also grateful for comments made by participants of an internal seminar at UNU-INTECH and another seminar at the Institute of Social Studies, The Hague, and especially Shulin Gu and Larry Rausch. Also grateful thanks to Lynn Mytelka, Sanjaya Lall, Thomas Hatzichronoglou, Ashok Desai, and Jeffrey James for their comments on an earlier draft. However, none of them are responsible for any errors that may still remain.

1. See European Commission (1997) for a full list of the 50 countries.
2. Though Kravis and Lipsey refer to their estimates as relating to the world as a whole, in fact they refer only to the market or OECD economies. See footnote "a" to table 1 in their study.
3. Lall's study is based on the UN COMTRADE dataset, but the level of aggregation is high at the three-digit level or higher in most cases. According to this study, high technology products include fine chemicals, pharmaceuticals, complex electrical and electronic machinery, aircraft, and precision instruments.
4. The COMTRADE database of the United Nations Statistics Division (UNSD) is by far the most comprehensive source of trade statistics as it covers over 110 reporting countries with time series starting from 1962 (SITC1), 1977 (SITC2) or 1988 (SITC3 correlated to Harmonized System (HS) codes), 1988 (HS 922) and 1996 (HS 1996). For details see http://unstats.un.org/unsd/comtrade/.
5. Hatzichronoglou (1997) explicitly states these three limitations.
6. To translate the Hatter–Davis industry classification into a definition of high technology trade, Braga and Yeats (1992) used the concordance between the SIC grouping and the SITC Revision 1 classification. In preparing the data on high technology trade, Braga and Yeats considered only SITC groups (at a four-digit level) that had a high technology weight above 50 per cent. Examples of high technology exports include aircraft, office machinery, pharmaceuticals, and scientific instruments. It is worth noting that this methodology rests on the somewhat unrealistic assumption that using United States input–output relations and trade patterns for high technology production does not introduce a bias in the classification (World Bank, 1999).
7. These nine countries together account for over 95 per cent of the world exports of high tech products.
8. However, Malta is one of the fastest growing countries in the world. Its per capita income has been growing at a rate of 4.5 per cent per annum since the mid-1970s. Examining how this country has become an important exporter of high tech products would be very instructive.
9. The World Bank has acknowledged me in its section on "Credits" in *World Development Indicators* (World Bank, 2000: 372).
10. For an excellent review of the literature on international technology transfer and the mechanics of "catching up", see Radosevic (1998).
11. This is computed by dividing a country's share of high tech exports in the world by its total share of world exports.
12. TPAC of the Georgia Institute of Technology. For details of the construction of the high technology indicators, see http://tpac.gcatt.gatech.edu/.
13. I do not have the data on high tech exports from Taiwan.
14. According the classification system followed in the USPTO, it divides patents into approximately 370 active classes. However, concordance tables between this classification system and the UN SITC are not available.
15. Revision 3 classes are not available. Therefore I am not in a position to classify the patent grants according to my definition of high technology.

REFERENCES

Braga, C. A. Primo and A. Yeats (1992) *How Minilateral Trading Arrangements May Affect the Post-Uruguay Round World*, Washington, D.C.: The World Bank, International Economics Department.

Davis, L. (1982) *Technology Intensity of U.S. Output and Trade*, Washington, D.C.: US Department of Commerce, International Trade Administration.

European Commission (1997) *Second European Report on Science and Technology Indicators*, Brussels: European Commission.

Griliches, Z. (1990) "Patent Statistics as Economic Indicators: A Survey", *Journal of Economic Literature* XXVIII: 1661–1707.

Hatter, V. L. (1985) *U.S. High Technology Trade and Competitiveness*, Office of Trade and Investment Analysis, Staff Report, Washington, D.C.: U.S. Department of Commerce, International Trade Administration.

Hatzichronoglou, T. (1997) *Revision of the High-Technology Sector and Product Classification*, OECDE/GD(97)216, Paris: OECD.

Huff, W. G. (1999) "Singapore's Economic Development: Four Lessons and Some Doubts", *Oxford Development Studies* 27(1): 33–55.

INTECH (2000) *Exports of High Technology Products from Developing Countries 1988–1998*, data on CD-ROM, Maastricht: United Nations University, Institute for New Technologies.

Kravis, I. B. and R. E. Lipsey (1992) "Sources of Competitiveness of the United States and Its Multinational Firms", *The Review of Economics and Statistics* LXXIV(2): 193–201.

Lall, S. (1997) "Policies for Industrial Competitiveness in Developing Countries: Learning from Asia", Oxford: Queen Elizabeth House, University of Oxford, unpublished.

—— (1998) "Exports of Manufactures by Developing Countries: Emerging Patterns of Trade and Location", *Oxford Review of Economic Policy* 14(2): 54–73.

Lall, S., M. Albaladejo and E. Aldaz (1999) "East Asian Exports, Competitiveness, Technological Structure, Strategies", Oxford: Queen Elizabeth House, University of Oxford, unpublished.

National Science Foundation (1995) *Asia's New High-Tech Competitors*, NSF 95-309, Arlington, Va.: National Science Foundation.

National Science and Technology Board (1988) *National Survey of R&D in Singapore*, Singapore: National Science and Technology Board.

OECD (1986) *OECD Science and Technology Indicators No. 2*, Paris: OECD.

Office of Technology Policy (1998) *The New Innovators, Global Patenting Trends in Five Sectors*, Washington, D.C.: U.S. Department of Commerce.

Pavitt, K. and L. Soete (1980) "Innovative Activities and Export Shares: Some Comparisons between Industries and Countries", in K. Pavitt, ed., *Technical Innovation and British Economic Performance*, London: Macmillan, pp. 38–66.

Radosevic, S. (1998) *International Technology Transfer and Catch-up in Economic Development*, Cheltenham (UK) and Northampton, Mass. (USA): Edward Elgar.

UNCTAD (1999) *World Investment Report 1999*, Geneva: United Nations.
UNIDO (1999) *International Yearbook of Industrial Statistics 1999*, Vienna: United Nations.
USPTO (1999) *Industrial Patent Activity in the United States*, part 1 on CD-ROM, Washington, D.C.: USPTO.
World Bank (1999) *World Development Indicators 1999*, on CD-ROM, Washington, D.C.: World Bank.
—— (2000) *World Development Indicators 2000*, Washington, D.C.: World Bank.
World Technology Evaluation Center (1997) "Electronics Manufacturing in the Pacific Rim", online at: http://itri.loyola.edu/em/toc.htm.

3
Development strategies and innovation policies in globalisation: The case of Singapore

Alexander Ebner

1 Introduction

Building institutional networks for the assimilation of technological innovations has emerged as an indispensable policy rationale in sustaining economic development, and providing opportunities for catch-up growth and structural upgrading in a global economy which has been shifting its technological foundations quite rapidly towards information and communication technologies. This type of developmental institution-building implies a strategic coupling of innovation policies and development strategies. Innovation policies may be perceived as specific policies for the promotion of local capabilities in the domain of innovation and technology assimilation, focusing on firms while accounting for institutional networks among private and public agents. These policies are framed by more comprehensive development strategies which denote those structural and institutional components that promote the realisation of a broad set of socio-economic goals, thus informing strategic efforts regarding policy coordination on a micro-, meso-, and macro-economic level. However, in dealing with the developmental challenges which are actually posed by globalisation, a complex set of technological, institutional, and spatial aspects needs to be considered.

Infrastructural investments and institutional adaptations are crucial elements of the development process, which may be assessed with respect to the evolution of specific techno-economic paradigms, perceived as ideal types of productive organisation (Dosi, Freeman and Fabiani, 1994: 11n). In coping with paradigm changes, then, systems of innovation denote those institutions within a territorial setting that take part in the generation and diffusion of innovations, combining components such as R&D facilities, education programmes, and patent systems. This institutional setting, as well as the specific industrial structures that shape interactive learning among users and producers, thus determines the local capacity for international competitiveness (Ebner, 1999: 143n). The impact of globalisation, however, points to the international restructuring of innovation systems, involving a drive towards commercialisation through transnational private–public partnerships.[1] Related to these efforts are attempts to build alliances between local authorities and transnational capital with an emphasis on the structural level of interaction (de la Mothe and Paquet, 1996: 26n).[2]

Indeed, technological globalisation is based on the geographically dispersed and locally specialised operations of multinational enterprises, integrated within corporate production and innovation networks that arrange technology flows among corporate centres and subsidiaries (Cantwell, 1999: 238). In this context, technology assimilation in developing economies should promote modification of established production structures, allowing for internal improvements and further spill-over effects which may stimulate local capabilities for innovation (Bell and Pavitt, 1993: 259n). Consequently, development strategies increasingly focus on the attraction of foreign direct investment, providing appropriate institutional conditions for industry-specific technology assimilation, as well as supporting the international expansion of local enterprises.[3] This involves continuous learning, although strategies of leapfrogging in the infrastructural domain seemingly succeeded in the East Asian economies (Hobday, 1995: 161). However, adaptive technological learning in late industrialisation may face stagnation as soon as the technology frontier is approached. Thus the local formation of sufficiently efficient innovation capabilities remains an indispensable factor in the orientation of innovation-based development strategies (Amsden and Hikino, 1993: 243n).

Accounting for these aspects, the present paper argues that the Singaporean development experience illustrates most convincingly the prospects as well as the limits of innovation policies and development strategies in the context of globalisation. Indeed, Singapore has been singled out as a successful example of industrial policies promoting development efforts by strategic intervention, especially in terms of fiscal and investment incentives, framed by public policies that aim at safeguarding the conditions for structural change (Soon and Tang, 1993: 18). In particular, the Singaporean development experience highlights the attraction of foreign direct investment as a means of employment creation and technological upgrading. This orientation has been set in relation to Singapore's regional position as a manufacturing and service hub, for Singapore is viewed as an infrastructural nodal point interconnected to transnational business networks, and thus adapting to the changing structural conditions of the world economy. Indeed, the Singaporean policies for locational competitiveness are based on a persistent structural adjustment, aiming at a pragmatic harmonisation of the needs of international investors with local developmental objectives (Ho, 1993: 47n). Consequently, reflecting the spatial dimension of technological globalisation, Singapore has been assessed as a role model for the development of an innovation-oriented metropolitan agglomeration (Castells, 1996: 390).

In order to sustain the competitive position as a trade, service, and communication hub, various infrastructural efforts have been launched, promoting the developmental vision of Singapore as an "intelligent island" in a global knowledge-based economy. However, obstacles in the realisation of the corresponding strategies may be due to problems of infrastructural and technological leapfrogging. The latter has been subject to further controversies on the internal causes of a seemingly less favourable performance of total factor productivity in the Singapore economy. However, this paper argues that an assessment of the Singaporean development trajectory still needs to account for explanatory factors underlying the macro-economic aggregates, thus favouring the institutional factors of the development process. This implies a reconsideration of the institutional embeddedness of technology assimilation, also pointing to the specific institutional foundations of innovation policies and development strategies in the Singapore economy.

Accordingly, the research questions guiding this paper are as follows: What are the specific development strategies and innovation policies that have shaped the performance of the Singapore economy? What is the role of innovation and technology assimilation in that performance, and what are its specific institutional foundations? In resolving these questions, this paper proceeds in two sections. First, the economic contours of the Singaporean development process are considered, accompanied by a discussion of current efforts in sustaining its dynamism by implementing innovation-oriented policies with a conceptual bias towards locational aspects. Second, the emerging Singaporean system of innovation is explored, with university–industry relations as well as the relationship between local government and international capital as essential characteristics. This leads to the conclusion that the changing techno-economic paradigm offers persistent challenges for a structural upgrading of the Singapore economy. Strategic problems then pinpoint the formation of institutions for the support of knowledge spill-over effects which should contribute to the evolution of local innovation capabilities.

2 Contours of the Singaporean development experience

2.1 The Singaporean trajectory of growth and development

The Southeast Asian city-state of Singapore, an island of about 3 million resident inhabitants at the southern tip of the Malayan peninsula, represents the smallest of the East Asian newly industrialising economies regarding territory and population. Singapore exhibits the highest GNP per capita in East Asia while occupying a very favourable position with regard to other development indicators (World Bank, 2000: 230n). Singapore's development from a formerly British colonial entrepôt trade port to the major business location in the present ASEAN area is due to various factors. It was established as a trade post by the British for logistical motives in an international context. A favourable geographical location meant it lay on major naval trade routes and has benefited from corresponding communication networks since the nineteenth century. Staple port activities related to trading fuelled the development process, laying the structural foundations for service and manufacturing activities (Huff, 1994: 7n).

The corresponding outward-oriented growth mode of a re-export economy regained its importance after Singapore's independence, followed by factual expulsion from the Malayan Federation in 1965. The economic success of export orientation, involving free trade policies and the attraction of foreign direct investment, reflected a favourable economic setting with expanding volumes of international trade and investment, benefiting all the Asia-Pacific economies (Islam and Chowdhury, 1997: 10n). Confronted with comparatively poor initial endowments in physical and human capital, the Singaporean government attempted to capitalise on infrastructural assets inherited from the British, primarily the port and its refinery equipment (Huff, 1994: 273n). Policies of infrastructural upgrading then encouraged incentives for international investors, thus supporting Singapore's emergence as a manufacturing location and service hub for multinational enterprises, including its role as a financial centre.[4]

Among the most striking features of Singapore's development since the 1960s are the high growth rates of per capita GDP, the outstanding role of exports in manufacturing, and the structural transformation towards manufacturing as well as financial and business services, accompanied by high savings and investment rates, low inflation rates, and macro-economic stability (Huff, 1994: 301n). The growth of aggregate output in the Singapore economy from 1965 to 2001 is indeed impressive, as depicted in table 3.1, illustrating a sustained growth path that was also characterised by strong recovery impulses in the aftermath of recessive

Table 3.1 Output and national income in Singapore, 1965–2001

Year	GNP (current market prices), $M	Per capita GNP (current market prices), $M	GDP (1990 market prices), $M
1965	3,052.0	1,618.0	7,718.0
1970	5,861.0	2,825.0	13,882.0
1975	13,566.0	5,903.0	22,329.0
1980	24,188.0	9,941.0	32,881.0
1985	40,330.4	14,740.6	45,344.9
1986	40,212.9	14,712.0	46,388.0
1987	43,140.6	15,547.0	50,899.9
1988	51,505.5	18,097.0	56,821.1
1989	59,736.0	20,381.5	62,288.8
1990	68,306.3	22,645.0	67,878.9
1991	75,328.2	24,378.8	72,860.9
1992	82,433.8	25,938.9	77,393.8
1993	93,466.0	28,675.8	85,473.2
1994	109,060.6	32,424.7	95,230.2
1995	121,483.3	35,021.9	102,808.1
1996	130,129.7	35,454.0	110,729.5
1997	149,450.4	39,394.0	120,190.8
1998	145,827.7	37,193.0	120,081.2
1999	143,507.2	36,323.0	128,404.5
2000	160,913.4	40,051.0	141,571.5
2001	154,644.5	37,433.0	138,682.6

Source: Department of Statistics (1996a: 2; 1996b: 4n; 2000: 2; 2001: 2).

phases. In the period under consideration, Singaporean GDP at 1990 market prices increased almost 18-fold, while per capita GNP at current market prices increased 23-fold. Accordingly, from the 1960s to the 1990s, average annual growth rates of GDP were 8–10 per cent, expressing sustained hyper growth of national income (Low et al., 1993: 30n).

Even recession phases in the early 1970s, the growth slowdown of the mid-1980s, and the Asian financial crisis in 1997 could not reverse this process, for on each occasion the growth performance soon regained its strong impetus. However, the dependence on foreign markets, especially the United States and Japan, has continuously exerted its influence on Singapore's export-oriented economy (Ministry of Trade and Industry, 2001: 14).

Moreover, gross fixed capital formation grew at an average annual rate of 11 per cent from 1965 to 1995, with brief phases of decline only during recessions in 1975, 1985, and 1997. This performance was paralleled by exceptionally high savings rates which increased from 11 per cent of GNP in 1965 to 50 per cent during the 1990s (Department of Statistics, 1996b:

3n, 2001: 2). This pattern of macro-economic growth variables may be related to government-sponsored saving mechanisms, channelled by the social security schemes of Singapore's Central Provident Fund since the 1960s. It has been argued that corresponding positions of fiscal surplus enabled the government to act as a public venture capital investor, involving investment abroad, whereas domestic investment efforts were backed by capital inflows in terms of foreign direct investment (Huff, 1994: 332n).

Supporting this role of the external sector, including the expansion of Singapore's foreign trade volume which has currently exceeded GNP threefold, a stability-oriented exchange rate policy has accompanied the established trade regime. Another aspect of macro-economic stability in conditioning economic growth has been provided by the low inflation rate, as reflected by the consumer price index. These factors seem to be at odds with the situation of labour shortage in the Singaporean workforce. However, they illustrate the result of the regulation of industrial relations and wage formation by the tripartite corporatist National Wages Council (Soon and Tang, 1993: 34n). Consequently, Singapore's growth trajectory, as performed by macro-economic indicators, rests upon specific institutional foundations which need to be associated with a pattern of governmental interventions, in the formation of factor prices, which has not blocked the market price system for goods and services.

As depicted in table 3.2, the rapid growth of the Singapore economy has been paralleled by structural changes in the composition and contribution of sectors and industries, most clearly related to the expansion of the manufacturing industries. These were continuously marked by the impact of electronics and chemical industries, with almost half the manufacturing industries' output and value addition provided by electronic products and components during the 1990s (Department of Statistics, 2000: 88n). The latter derived their growth dynamism principally from international demand for semiconductors and computer peripherals, confirming Singapore's role as a manufacturing hub in the transnational business networks of the electronics industries. However, adding to the expansion of financial and business services which have come to dominate the service sector, almost 70 per cent of Singapore's GDP is made up of service activities in a broad sense (Ministry of Trade and Industry, 2001: 14n). The service sector has an equally high employment share in total workforce allocation, with a majority of the workforce absorbed by local small and medium enterprises which tend to exhibit comparatively low productivity levels (Ministry of Trade and Industry, 2001: 21n, 61).

Accordingly, the decisive impulses for Singaporean growth and development have been provided by the manufacturing and service activities of multinational enterprises, which have replaced the staple port economy

Table 3.2 Structural change in the Singapore economy, 1965-1995

	GDP by industry (current market prices), percentage distribution						
Year	Manu-facturing	Utilities	Con-struction	Com-merce	Transport and com-munication	Financial and business services	Other services
1965	15.2	2.2	6.5	27.2	11.5	16.6	17.6
1970	20.2	2.6	6.9	27.4	10.7	16.6	12.9
1975	23.3	1.8	7.9	24.8	10.8	18.2	11.1
1980	28.1	2.1	6.2	20.8	13.5	19.0	8.8
1985	22.0	1.9	10.0	15.9	12.6	25.6	11.0
1990	27.2	1.8	5.2	17.9	12.2	25.0	10.3
1995	24.9	1.5	6.7	18.6	11.1	26.9	10.0

Source: Department of Statistics (1996b: 8n).

with its trade-related services (Huff, 1994: 299n). Indeed, the outward-oriented attraction of foreign direct investment through the provision of institutional and infrastructural resources, in terms of both physical and knowledge infrastructures, has been a key concern in Singaporean development strategies, allowing for a shift from labour-intensive to capital-intensive industries, which are currently oriented towards knowledge-intensive activities. The operations of multinational enterprises thus have been a major factor in the evolution of local technological capabilities (Wong, 1999: 272n).

The outstanding impact of foreign direct investment in the manufacturing sector of the Singapore economy is depicted in table 3.3, focusing on foreign and local net investment commitments in the manufacturing sector. Since the 1970s, both the foreign and local sectors have experienced a massive expansion of net investment commitments in manufacturing industries, with the foreign sector accounting for about 70 per cent of the total investment commitments in manufacturing. A gradually increasing role of local enterprises has been registered, as their relative contributions to total investment commitments seemed to suffer temporarily from the aftermath of the 1985 recession, yet experienced a steady expansion through the 1990s. However, an assessment of local enterprises and their investment performance needs to account for government-related local enterprises as a representation of the specific Singaporean blend of governmental entrepreneurship (Vogel, 1991: 80n).

Concerning the origin of foreign investment resources, Japanese firms have taken the top position in cumulated foreign equity investment during the 1990s, reflecting the regional integration of the East and South-

Table 3.3 Foreign and local net investment commitments in Singapore's manufacturing industries, 1972–1999

Year	Foreign ($M)	Local ($M)	Total ($M)	Local/total (%)
1972	156.3	38.2	194.5	19.6
1980	1,199.0	218.9	1,417.9	15.4
1985	888.0	232.4	1,120.4	20.7
1986	1,191.0	259.0	1,120.0	23.1
1987	1,448.0	295.0	1,450.0	20.3
1988	1,658.0	350.0	1,743.0	20.0
1989	1,625.0	333.0	2,007.0	16.6
1990	2,217.9	269.5	2,487.4	10.7
1991	2,461.1	472.9	2,934.0	16.1
1992	2,733.0	748.0	3,481.0	21.4
1993	3,177.1	745.5	3,922.6	19.0
1994	4,327.4	1,437.2	5,764.6	24.9
1995	4,852.4	1,956.7	6,809.1	28.7
1996	5,791.8	2,293.3	8,085.1	28.3
1997	5,963.8	2,524.6	8,488.4	29.7
1998	5,257.1	2,615.9	7,829.4	33.4
1999	6,257.1	1,780.3	8,037.4	22.1

Sources: Department of Statistics (1996a: 96; 2000: 95), Low et al. (1993: 361).

east Asian economies. There is also marked expansion of foreign equity investment in financial services, denoting a segment that parallels foreign investment in manufacturing, quite in agreement with Singapore's global city status (Department of Statistics, 1996b: 14). Still, the United States has remained the decisive country of origin for foreign net investment commitments in manufacturing, providing almost 50 per cent during the 1990s (Department of Statistics, 2000: 71, 95). This pattern is in agreement with the macro-regional structure of the world economy, as well as with Singapore's orientation towards electronics industries.

Nonetheless, outward-oriented industrialisation has not yet produced a coherent internal structure of economic activities, for the productivity performance of a majority of local small and medium enterprises, especially in the local service sector, fails to meet international standards concerning cost and price competitiveness, thus pointing to a persistent lack of local technological capabilities (Yun, 1998: 398n). It may be suggested that those local firms which interact with multinational enterprises as suppliers and subcontractors are those that benefit most decisively from technology assimilation in terms of sustained learning effects. Indeed, this assessment has been corroborated by empirical observations in the Singapore electronics industry (Hobday, 1995: 136n). The development strategies of the Singaporean government therefore delegate a crucial role to local learning effects as a strategic element in the restructuring of the Singapore economy, which has been challenged again by a changing techno-economic paradigm.

2.2 Development strategies for economic growth and international competitiveness

After a brief phase of import substitution within the Malayan Federation, lasting until 1965, Singapore attempted to combine the autonomy of a nation-state with the locational advantages of an infrastructurally well-endowed city-state. Hence, soon after attaining national independence, Singapore promoted the policy of locational competitiveness, aimed at attracting foreign direct investment by providing an institutional and infrastructural environment conforming with the needs of international investors.[5] Attracting labour-intensive foreign direct investment as a means of creating employment has dominated the policy rationale since the mid-1960s, moving gradually to a more capital-intensive investment pattern in the mid-1970s, which implied a shift in the pattern of incentives provided by the Singapore authorities (Rodan, 1989). In the aftermath of the recession of 1985, and also due to the emergence of locational competitors in the newly industrialising economies of the ASEAN area and beyond, Singapore's government turned to a strategy of providing opportunities for high value-added business activities, primarily knowledge-intensive segments that should preserve Singapore's role as a service and manufacturing hub (Islam and Chowdhury, 1997: 206n).

The complex array of economic policies associated with that developmental strategy has been designed as a flexible and pragmatic approach towards maintaining competitiveness in support of a continuous upgrading of the value-added operations of the subsidiaries of multinational enterprises, and local enterprises interacting with them, involving spill-over effects as well as self-directed benefits. Indeed, not only structural policies, but also monetary and fiscal policies have been subordinated to the primacy of maintaining locational competitiveness. For instance, the government-guided corporatist tripartite arrangement that regulates nominal wages and governs industrial relations within the National Wages Council has been perceived as a locational policy instrument (Chew and Chew, 1996).

The intensity of regulation efforts differs with regard to the relevant production factors. While the immobile factor land and the comparatively immobile factor labour have been subject to intense regulative interventions, the mobile factor capital has been subject to measures with a rather indicative character, involving a broad range of material incentives. This incentive-based approach has allegedly worked as an effective mode of indicative development planning (Huff, 1994: 339n). The lack of rent-seeking behaviour in the determination of market interventions seems to have contributed most successfully to that policy (Bercuson, 1995: 18n). Nonetheless, Singapore's government remains a most influential economic agent, as it is continuously intervening in the business sector, despite ongoing privatisation of state enterprises and government-related enterprises which face liberalised markets, most prominently in the telecommunications sector.[6]

In addition to the facilitative activities of government regarding the provision of institutional and infrastructural resources that are conducive to the market process, state-owned enterprises have played a decisive role as institutional supporters of entrepreneurship.[7] This situation has been held responsible for excluding local entrepreneurship, basically overseas Chinese business networks (Huff, 1994: 320n). Furthermore, Singaporean government boards have exercised entrepreneurial functions directly by stimulating economic restructuring and technological change, thus introducing routine-breaking innovations in the economic sphere. These procedures represent a specific blend of neo-mercantilist development planning and market-oriented policies (Ebner, 2004).

The Economic Development Board (EDB) may be perceived as the core organisation in Singapore's policy networks, exercising a leading function in the hierarchy of government boards. The EDB has been continuously guiding Singapore's development trajectory through "strategic pragmatism", that is the practice of matching the strategic interests of the government with the corporate needs of international investors which are subject to the volatile conditions of global markets (Schein, 1996). Indeed, the EDB was primarily concerned with the provision of industrial estates in the 1960s, evolving as a major organisation in Singapore's development planning in the 1970s, then heralding the restructuring efforts of the 1980s towards higher value-added operations in targeted industries. This implied a broader approach to service provision in terms of cooperative interactions with multinational enterprises as a rationale of locational policies during the 1990s, coordinated by the EDB with other government boards and agencies (Low et al., 1993: 63n).[8]

The pattern underlying these activities seems to provide evidence for ongoing learning efforts regarding policy orientation and implementation. Singapore actually experienced cases of government failure which had to be corrected immediately. The high-wage policy in the early 1980s, for instance, was originally initiated in order to support the relocation of labour-intensive manufacturing, with the aim of enforcing a local expansion of high value-added production segments with sustained productivity and income effects. However, this wage policy for the promotion of factor substitution was confronted with an actual shortage of skilled manpower, accompanied by an unwillingness of foreign investors to comply with the wage constraints imposed by the government. All of this contributed to an economic downturn in 1985 (Yun, 1998: 381n). The failure of this strategy, which had to be cancelled in the aftermath of that brief recession, marked a policy reorientation towards technological upgrading and structural competitiveness by means of distinct innovation policies that should also reinforce the participation of foreign investors in local policy networks. These aspects were essentially formulated in a government report on policies for maintaining Singapore's international competitiveness (Ministry of Trade and Industry, 1986).

This outward-oriented responsiveness is well illustrated by the fact that Singapore's development strategies have accounted for the restructuring of multinational business networks through the strategic perception of localisation and regionalisation as parallel processes in the context of globalisation. In addition to the establishment of Singapore as a business and communication hub, an attempt was made to assist multinational enterprises actively in their relocation efforts throughout East and Southeast Asia (Low et al., 1993: 157n). Regionalisation as a policy focus has been also addressed by the "growth triangle" strategy, that is an effort to establish special zones for labour-intensive manufacturing processes in Johor and Riau, border regions of Singapore's land- and labour-affluent neighbours Malaysia and Indonesia. Local enterprises, however, should be supported in building up ventures overseas, focusing on ASEAN, India, and China, again supporting Singapore's role as a regional hub. It has been argued that this kind of responsiveness mirrors a continuous "reinvention" of governance mechanisms, assessed as a key indicator of Singaporean state capacity in governing the economy (Low, 1998: 269n).

The post-recession orientation of the late 1980s pinpointed strategies for industrial restructuring that were dedicated to total factor productivity, combining it with structuralist perspectives on competitiveness.[9] An important influence on this plan was exerted by the cluster approach, as formulated by Michael Porter, which underlines the systemic nature of institutional conditions and industrial linkages that constitute the clusters of competitive industries. According to the related development stages model, an economy usually passes, first, a stage of factor-driven growth with comparative advantages in factor endowments and extensive use of these factors. Investment-driven growth, the second stage, with investment and foreign technology fuelling the growth process, is followed by the third stage of innovation-driven growth, whereas a final wealth-driven stage signals stagnation (Porter, 1990: 545n). Porter suggested that Singapore performs at the factor-driven stage, lacking "home base" qualities for the development of competitive local industries (Porter,

1990: 566). Lessons for developing economies, also of relevance for Singapore, underlined the formation of industrial clusters, advanced R&D capabilities, appropriate education systems, upgraded infrastructures, the internationalisation of domestic structures, and the nurturing of local firms to become multinational enterprises (Porter, 1990: 676n).

These conceptual positions have guided Singapore's restructuring efforts since the late 1980s. The government report on Singapore's competitiveness, which responded to the mid-1980s recession, dealt primarily with cost competitiveness, focusing on wage regulation, whereas long-term development was said to be based on Singapore's role as a business and service hub. In this framework, the expansion of R&D activities was reconsidered with regard to tax incentives and funding schemes, accompanied by manpower provision and a distinct R&D infrastructure (Ministry of Trade and Industry, 1986: 147n). Subsequent government proposals such as the "Strategic Economic Plan", published in 1991, reiterated these motives. This policy programme aimed at attaining the status of a developed economy by the year 2030, focusing on the realisation of an innovation-driven development pattern in the sense defined by Porter. This should imply the expansion of institutional networks for innovation, based on the infrastructures of a knowledge-intensive techno-economic paradigm. Strategic thrusts included the enhancing of human resources in terms of education and training, the coordination of tripartite regulation with key economic interest groups, and an intensified international orientation regarding knowledge infrastructures and networks, as well as the promotion of an innovative milieu containing skilled manpower and R&D facilities (Economic Planning Committee and Ministry of Trade and Industry, 1991: 57n).

These efforts should be accompanied by specific industrial strategies, namely the development of manufacturing and service clusters, as well as the redevelopment of local enterprises, with the aim of productivity enhancement. Economic resilience in carrying out these strategies should be addressed by the maintenance of international competitiveness through flexible wage formation and other institutional arrangements in the domain of unit labour costs, productivity, and profitability, accompanied by efforts in structural diversification regarding "home base" qualities for multinational and local enterprises alike. Indeed, multinational enterprises should be viewed as long-term partners in promoting economic development (Economic Planning Committee and Ministry of Trade and Industry, 1991: 71n). All of this was translated into assistance schemes for cluster development, innovation support, business service incentives, and schemes for the funding of overseas infrastructural projects, administered by the Economic Development Board (1996).[10] A related cluster development initiative was launched in 1999 under the label of the "Industry 21" plan, with information technology and life sciences as prominent areas of concern (Economic Development Board, 2000: 10n).

Moreover, in agreement with the notion of an emerging techno-economic paradigm based on information technologies, Singapore has been set to become an "intelligent island", based on the design of a competitive national information infrastructure (National Computer Board, 1992). Related policies have been marked by an emphasis on liberalisation and privatisation, paralleled by a refinement of infrastructural facilities and services, as illustrated by the "Singapore One" initiative in establishing an economy-wide broadband network. This coincides with operations in strategic domains such as e-commerce and e-government (Ministry of Trade and Industry, 2001: 78n). All of these activities hint at a decisive shift in Singapore's development strategies which also affected the domain of innovation policies, perceived as policies for supporting the local assimilation of new technologies. Examining the institutional foundations of that process implies an exploration of the Singaporean system of innovation.

3 Evolution of the Singaporean system of innovation

3.1 Innovation policies for technology assimilation

The Singaporean system of innovation may be portrayed as an institutional network that co-evolves with its technological and spatial environment, combining the policy capacity of a national government with the metropolitan features of a global city. In particular, its institutional networks include multinational enterprises, public enterprises, government-related enterprises, private local enterprises, and government boards and agencies, as well as research and education facilities. However, due to the structural dominance of multinational enterprises, the Singaporean innovation system has been perceived as a model that leverages on foreign direct investment, focusing on the facilitation of technological learning in local enterprises, as induced by foreign direct investment (Wong, 1995: 20n).[11] Accordingly, the subsidiaries of multinational enterprises dominate R&D activities in the private sector, accompanied by the related activities of their local suppliers as well as by strategic operations performed by government-linked companies and public statutory boards.

However, apart from the position that multinational enterprises need to be viewed as strategic partners in economic development, thus representing a constitutive component of the Singapore economy, the structural specificity of local enterprises in the manufacturing industries needs to be underlined. This is most important with regard to local efforts in R&D, which have been singled out as an important domain of policy intervention. R&D activities are carried out by local firms such as small- and medium-sized enterprises in segment supporting multinationals, government-linked companies in technology-intensive industries, and private sector start-up firms in promising technology sectors (Wong, 1995: 15n). This pattern implies that appropriate policies need to account for an actually existing structural diversity of innovation activities. Consequently, user–producer relations in the generation of innovation have been also dominated by multinational enterprises. It has been maintained that they have insufficiently affected the performance of small- and medium-sized enterprises in the service and manufacturing sectors, at least during the early phases of Singaporean catch-up growth.[12]

Therefore, Singaporean policies for building an innovation system are concerned with establishing bridging institutions that facilitate knowledge spill-over and learning effects, thus combining the rationale of international competitiveness with a cooperative approach to skills development and the technological upgrading of local enterprises. The structural dimension of the Singaporean innovation system, with its specific linkages, is complemented by an institutional pattern in support of technology assimilation. In general, this consists of Singapore's education sector, with its two universities and four polytechnics, accompanied by public research facilities that are also associated with the cooperative efforts of multinational enterprises. Moreover, industrial training facilities have been set up in cooperation with international partners, addressing foreign enterprises as well as foreign government agencies. This underlines again the outward-oriented character of the Singaporean innovation system, although an extended interaction between private sector enterprises and public research facilities on an international scale has become a prominent policy topic only since the mid-1990s (Ebner, 2004).

Singapore's government shapes R&D activities and further innovation policy measures through its National Science and Technology Board (NSTB), while the EDB, as a hegemonic government board in terms of strategic relationships with foreign investors, exercises primarily an indirect influence by moulding the policy context, which is of course also affected by other government boards and ministries. Still, investment schemes administered by the EDB also promote innovation activities. However, during the 1980s, technological innovation was of minor importance, as Singapore's development orientation was originally based on the provision of manpower and infrastructure for standard manufacturing and service operations, yet with a continuous concern for increasing productivity levels. It is noteworthy that the Ministry of Technology was disbanded during a reorganisation phase in 1981, aimed at increasing administrative efficiency, and government boards were established instead. The recession of the mid-1980s and subsequent policy debates then heralded the emergence of a distinct innovation policy, leading to the establishment of the NSTB and the formulation of the National Technology Plan in 1991. This plan announced the promotion of industrial R&D, perceived in terms of private sector activities concerning applied innovation activities. Related measures should include the selective funding of R&D, as well as the provision of manpower and institutional infrastructures (National Science and Technology Board, 1991).

The essence of the Singaporean approach to building a national innovation system is well described by the notions of service provision and market orientation, quite in accordance with the official goal of transforming Singapore into a technologically advanced service hub, including the provision of R&D facilities. Indeed, the Singaporean system of innovation has been bound to complete a "total business centre" strategy (Economic Planning Committee and Ministry of Trade and Industry, 1991). Schemes for attracting high value-added business headquarter functions have been paralleled by the implementation of research grants and fiscal incentives for the support of innovation activities, benefiting companies such as Hewlett-Packard and Philips which have set up their regional R&D centres in Singapore and greatly expanded their local production and service capacities, yet also affecting local firms in their expanding innovation efforts. In this domain, grants such as the Research Incentive Scheme for Companies and the Innovation Development Scheme accompany fiscal schemes for the tax deduction of R&D expenditure. However, another decisive component in that programme, next to the important aspect of physical infrastructure, hinted at the education and training of manpower which should meet the requirements of an increasingly knowledge-intensive type of production (Mani, 2000: 25n).

Still, a lack of basic research and a shortage of skilled manpower in science and technology remain critical features of the Singaporean innovation system. In the mid-1980s, the government explicitly decided to avoid supporting basic research due to both the uncertainty concerning returns on investment and financial risks, as well as the lack of adequate manpower (Ministry of Trade and Industry, 1986). This argument should also account for the investment behaviour of multinational enterprises with their tendency to adjust R&D towards local knowledge agglomerations. Moreover, it should mirror related aspects of the Japanese and Taiwanese systems of innovation, perceived as benchmarking models, with their original focus on applied research in contrast to the strong performance of basic research in the United States. Singapore's emphasis on the commercial segments of scientific and technological activities, prepared in cooperation with multinational enterprises, thus has neglected the need to develop local capabilities in basic research as an indispensable component of a sustainable innovation portfolio. This is reflected by a low level of interaction between established public research facilities and private enterprises, accompanied by a lack of networking interaction among local small- and medium-sized enterprises (Wong, 1995: 34n). Assessing these arguments implies an empirical reconsideration of the Singaporean innovation performance.

3.2 The innovation performance of the Singapore economy

R&D activities are usually appreciated as activities which add high value and thus they play a prominent role in Singaporean innovation policies. Still, innovation processes are also of an informal and tacit nature, especially in small- and medium-sized enterprises, and are not altogether obvious when focusing on R&D as an input indicator of innovation. Moreover, output indicators, such as granted patents, are not of primary relevance in the Singaporean policy rationale, although they have recently been approaching the Hong Kong level of activity, as far as patents granted in the United States are concerned (Mani, 2000: 12). However, the level of expenditure on R&D as well as employed R&D manpower may serve as a useful approximation of the complexity of innovation activities in the Singapore economy. Table 3.4 depicts the development of R&D activities: Singaporean efforts in expanding R&D operations are obvious. Gross expenditure on R&D (GERD) increased more than 30-fold between 1981 and 1999. They continued to grow even in the midst of the Asian crisis, reaching $2,492 million in 1998 and $2,656 million in 1999. This corresponds with the fact that the GERD to GDP ratio grew from 0.26 per cent in 1981 to 1.84 per cent in 1999 (National Science and Technology Board, 2000: 3).

Nonetheless the Singaporean system of innovation still exhibits a "GERD lag" as compared with the other East Asian newly industrialised economies, not to mention the Japanese innovation system (Masuyama, 1997: 3n). In spite of a rapid growth performance, Singapore's GERD to GDP ratio still indicates a policy failure when the Singapore government's original ambition of achieving a 2 per cent ratio by 1995 is taken into consideration, although these target numbers primarily serve as rather symbolic yardsticks (National Science and Technology Board, 1991). Currently, these numbers have been projected more realistically, thus the modified target level of the GERD to GDP ratio of 1.6 per cent for the year 2000 has been attained earlier than projected. The corresponding goals of a ratio of 40 research scientists and engineers (RSEs) per 10,000 labour force as well as a minimum share of 50 per cent private

Table 3.4 Innovation input indicators in Singapore, 1981–1999

Year	Gross expenditure on R&D ($M)	GERD/GDP ratio (%)	Research scientists and engineers per 10,000 labour force
1981	81.0	0.26	10.6
1990	571.7	0.86	27.7
1991	756.7	1.02	33.6
1992	949.3	1.19	39.8
1993	998.2	1.07	40.5
1994	1,175.0	1.10	41.9
1995	1,366.6	1.16	47.7
1996	1,792.1	1.39	56.3
1997	2,104.6	1.50	60.2
1998	2,492.3	1.80	65.5
1999	2,656.3	1.84	69.9

Sources: Ministry of Trade and Industry (1997), National Science and Technology Board (2000: 3n).

sector R&D expenditure have already been met in 1995. Even the modified projected level of 65 RSEs per 10,000 labour force, originally calculated for the year 2000, was achieved in 1998 (National Science and Technology Board, 2000). However, all international benchmarking positions are inadequate, given Singapore's city-state status and its role as a metropolitan agglomeration. Benchmarking efforts with reference to global cities and metropolitan regions would provide more suitable material for a comparative assessment.

With respect to table 3.5, depicting sectoral contributions to R&D expenditure, it becomes obvious that firms in the private sector, mostly multinational enterprises, take the lead position, accounting for two-thirds of gross expenditure. Indeed, between 1990 and 1999, private sector R&D expenditure increased fivefold, from 309 million Singapore Dollars to 1670 million Singapore Dollars (National Science and Technology Board, 2000: 9). In 1999, private sector R&D spending concentrated on engineering sciences, focusing on the electronics, electrical, and mechanical segments, together with computer and related sciences, as well as chemical science. This reflects the impact of the lead industries in the manufacturing sector, and thus the proximity of the dominant pattern of R&D to the needs of industrial production. R&D expenditures in the public domain, including government, higher education, and public research facilities, also seem to follow this pattern, shaped by the industry-specific articulation of private sector concerns for the commercial application of new knowledge. Accordingly, the bulk of R&D expendi-

Table 3.5 Expenditure on R&D by research area, 1999

Research area	Private sector ($M)	Higher education sector ($M)	Government sector ($M)	Public research institutes ($M)	Total ($M)
Agricultural sciences	3.97	1.17	1.91	19.08	26.13
Computer and related sciences	395.81	29.20	56.48	58.54	540.03
Engineering	896.14	154.76	62.74	149.83	1,263.48
Aeronautical	5.06	0.00	6.53	0.00	11.58
Marine	9.87	1.77	6.35	0.00	17.99
Mechanical	298.29	34.40	16.48	18.17	367.34
Biomedical	2.88	6.33	0.00	0.00	9.21
Civil and architecture	3.35	33.85	1.98	0.00	39.19
Electrical and electronic	479.35	47.43	22.17	90.64	639.59
Material sciences and chemical	83.31	28.01	9.23	41.03	161.58
Metallurgy and metal	14.02	2.96	0.00	0.00	16.99
Biomedical sciences	34.22	46.79	29.05	3.62	113.68
Natural sciences	106.94	54.02	23.55	83.36	267.87
Biological	7.10	19.55	8.25	59.99	94.89
Chemical	98.10	12.73	6.98	0.94	118.76
Earth and related environmental sciences	1.42	0.60	0.85	20.02	22.88
Physical sciences and mathematics	0.32	21.15	7.48	2.41	31.35
Other areas	233.78	24.03	131.12	56.18	445.11
Total R&D expenditure	1,670.86	309.97	304.85	370.61	2,656.30

Source: National Science and Technology Board (2000: 11).

ture is concerned with experimental development and applied research, amounting to 86 per cent of total expenditure, whereas the pure and strategic types of basic research amount to 14 per cent, with a comparatively strong presence in public research institutes and centres, as well as in higher education (National Science and Technology Board, 2000: 10).

Table 3.6 depicts sectoral R&D expenditure contributed by local and foreign companies in 1999.[13] According to these data, multinational enterprises take the lead in providing the Singaporean innovation system with R&D impulses. This is immensely relevant in those industries which belong to the technologically most advanced production segments. Thus,

Table 3.6 R&D expenditures of local and foreign companies by industry, 1999

Industry group	(1) Foreign companies ($M)	(2) Local companies ($M)	(3) Total ($M)	(2)/(3) (%)
Manufacturing	780.99	585.05	1,366.05	42
Electronics	518.32	237.80	756.12	31
Chemicals	99.89	26.69	126.58	20
Engineering	102.41	237.73	340.14	69
Precision engineering	84.13	191.93	276.05	69
Process engineering	6.02	3.72	9.74	38
Transport engineering	12.26	42.09	54.35	77
Life sciences	58.55	31.13	89.68	34
Light industries/other manufacturing	1.82	51.71	53.52	96
Services	151.84	152.98	304.82	50
IT/communications	71.17	105.70	176.87	59
Finance and business	18.98	12.25	31.23	38
Other services	61.69	35.03	96.72	36
All industry groups	932.83	738.03	1,670.86	44

Sources: National Science and Technology Board (2000: 34), calculations by author.
Note: Definition of local: 30% or more locally owned; definition of foreign: less than 30% locally owned.

the local content of R&D activities amounts to only 31 per cent in electronics and 20 per cent in chemicals, well below the average 42 per cent level of the whole manufacturing sector.

Moreover, it is noteworthy that the IT-related services segment is still dominated by local contributions with a 59 per cent share, the highest local industrial share apart from the transport industry. Indeed, this service area is most important for the "intelligent island" development strategy which underlines the necessity of international co-investment and concentrated R&D efforts. Thus, regarding the developmental importance of information and communication technologies in the hierarchy of government priorities, and reconsidering the sustained efforts in mobilising international partners for upgrading Singapore's national information infrastructure, the actual representation of foreign investors seems to be rather weak. In this case, R&D profiles may also illustrate the strategic limits of building an innovation system with a high degree of outward orientation, especially in a domain with exceptionally high initial investment volumes. The strategies and expectations of multinational enterprises do not necessarily meet host country needs, or at least the strategies of the host country government. Still, apart from the design of a national information infrastructure, it seems that Singapore's innovation policies have accounted for this fact more effectively, as exemplified by the evolution of cooperative research, education, and training facilities. Sustained international contributions to institution building in this domain of knowledge generation and dissemination have actually assisted in establishing the institutional core of the Singaporean innovation system.

3.3 Industry–university relations, education, and skills formation

Reflecting the spatial dynamism of globalisation, Singapore's character as a regional manufacturing, trade, and service hub in Southeast Asia serves as the basis for realising the developmental goal of turning the city-state into a knowledge-intensive agglomeration, including academic centres of excellence which should have access to the global innovation networks of multinational enterprises with their high-level knowledge flows. Accordingly, establishing a local science base becomes crucial for sustaining Singapore's locational competitiveness under the conditions of technological globalisation.[14] The evolving science base of Singapore's innovation system is founded on interactions among private and public as well as local and international agents, with an exposed role for local tertiary education institutions and their affiliated research institutes. Indeed, Singapore's innovation policies have recently promoted the establishment of R&D facilities in the education sector, providing a knowledge infrastructure for R&D collaborations with multinational corporate partners, as well as supporting local entrepreneurial spin-off ventures in knowledge-intensive industries. Beyond the confines of innovation policy, therefore, these research institutes and centres represent an institutional cornerstone in Singapore's currently implemented development strategies (National Science and Technology Board, 1999).

What does this exposed position of the academic sector imply in analytical terms? Regarding the evolution of knowledge-based agglomerations with a global reach, the outstanding role of universities and research districts in driving the agglomeration of entrepreneurial ventures has been exemplified by the cases of Silicon Valley, with nearby Stanford University, and the Route 128 district around Boston, with Harvard and MIT as neighbouring academic centres of excellence (Saxenian, 1994). On a conceptual level, then, efforts to establish industry–university relations within the Singaporean innovation system may be interpreted in terms of the set-up of a "competence block", denoting agents and organisations that are involved in the creation of increasing returns to innovative research. This points to the core functions of universities and science parks in the evolution of an innovation system, for they provide filters and institutional bridges from scientific research to the market sphere, while additionally supplying skilled personnel for the domains of both research and entrepreneurship (Eliasson, 2000: 236n).[15]

This fundamental role of tertiary academic institutions in Singapore's development strategy may be exemplified with reference to the formation of an R&D cluster in the "technology corridor", set up in the southwest of the island. This corridor is aimed at making structural as well as institutional connections between the Jurong industrial estates, prominent as a location for the manufacturing operations of multinational enterprises, and the surrounding universities and polytechnics, accompanied by neighbouring business, service, and science parks. The corridor is thus meant as an agglomeration of knowledge-based agents from the private and public sectors, operationally biased towards R&D activities. The expansion of the science park located next to the main campus of the National University was announced as a first step in developing the corridor, referring to the pioneering establishment of science park facilities in 1980 (Economic Development Board, 1996). Indeed, illustrating the impact of that strategy, a second district belonging to the science park project was established in a nearby area in the late 1990s, with further expansion of its facilities under way. All of this represents the institutional realisation of an evolving competence block within Singapore's innovation system.

In this context, the National University of Singapore (NUS) is not only the leading academic organisation in higher education, but also a decisive player in university–industry collaboration on applied R&D. In order to fulfil its obligations, NUS provides a campus location for regionally outstanding research institutes, such as the Institute for Systems Science, the Institute for Molecular and Cell Biology, and the Institute for Microelectronics, indicating a research focus on information and communication technology, microelectronics, and life sciences. Reflecting the strategic bias towards international private–public partnerships, these institutes cooperate with multinational enterprises and international research partners. Corresponding modes of collaboration range from contract research to mutual agreements of understanding with partner companies such as IBM and Sony (National University of Singapore, 1996).

The second major academic player is Nanyang Technological University (NTU) where research is performed by institutes such as the Advanced Materials Research Centre and the Gintic Institute of Manufacturing Technology. Compared with NUS, R&D activities at NTU seem to have an even more applied orientation, closer to the innovation segment of product improvement (Nanyang Technological University, 1996). This assessment holds also for Singapore's four polytechnics, which are engaged in R&D activities only on a comparatively minor scale: Ngee Ann,

Singapore Polytechnic, Nanyang, and Temasek. Despite their focus on the support of local enterprises, international collaboration also remains relevant, as illustrated by the Japanese–Singapore Institute of Software Technology, set up by the Singapore polytechnics as a joint bilateral effort in applied research.

Nonetheless, despite these complex activities, it has been argued that the pattern of institutional interaction in the domain of industry–university relations needs to be singled out for further improvements, especially with regard to the integration of those small- and medium-sized enterprises in the local manufacturing and service industries which do not belong to the high tech segment of the private sector (Wong, 1999: 281n). Indeed, it seems that the spill-over effects from the knowledge-intensive high technology industries all over the Singapore economy, with its large segment of small enterprises in the service sector, have been rather insignificant so far, at least with regard to the corresponding productivity profiles. At this point, the logic of technological globalisation, with its combination of local specialisation and global flexibility, may contradict integrative development strategies which tend to seek a strengthening of structural cohesion in the local economy. However, the problem of comprehensive restructuring towards a knowledge-based economy involves not only knowledge-intensive industries and their specific linkages on a local and international scale, but also the provision of adequate manpower.

Indeed, the developmental role of higher education is not confined to the provision of R&D facilities. With an immense scarcity of both capital and land, labour has been Singapore's only readily available local resource. Thus, education and training of the workforce have been a fundamental component of the Singaporean development trajectory ever since the 1960s.[16] The formation of specialised highly skilled manpower depends on Singapore's selective tertiary education system, in which the local academic élite has been traditionally educated at the National University with its focus on advanced academic research. This academic tradition had traditionally neglected engineering as an academic discipline. However, the engineering sciences have been identified as a decisive element for local skills development, and engineering enrolment numbers have been continuously increasing (Department of Statistics, 2000: 232). In contrast to the intellectual style of Singapore's National University, Nanyang Technological University offers a more work practice-oriented education which focuses on engineering and business administration. A broad area of technological education is also supplied by the four polytechnics with their practice-oriented mission. In spite of these efforts, it has been proposed that the steady provision of local manpower for technologically advanced industries remains a critical factor in the evolution of the Singaporean innovation system (Wong and Ng, 1997: 139). This applies not only to knowledge-intensive industries with their specific workforce segments, but also to those comparatively labour-intensive industries in the manufacturing and service sectors which are subject to diverse skills upgrading schemes.

Therefore, an assessment of Singapore's innovation policies and development strategies also needs to account for the continual provision of workforce training programmes, which has been interpreted as a positive feature in all the East Asian newly industrialising economies, for skills upgrading most effectively supports firm-specific learning processes and thus allows for sustained efforts in technology assimilation (Hobday, 1995). A strategy of enforcing productive discipline informed the related labour policies in the 1960s and early 1970s, then shaped by political struggles between Singapore's single party government and an opposition of militant trade unions. However, since the 1970s the emphasis of these policies has shifted towards the promotion of productivity enhancement by skills upgrading, which should also legitimise increasing real wages, while labour-intensive manufacturing activities have been gradually relocated to neighbouring low-wage areas in the Asia-Pacific area (Rodan, 1989). Completing the shift from mere workforce adaptation to labour-intensive manufacturing operations in local subsidiaries of multinational enterprises towards the developmental goal of achieving international standards in high value-added skill segments on an economy-wide scale, then, the orientation of these schemes reflects a persistent concern with Singapore's locational competitiveness.

The most important skills development programmes are implemented by the Singapore Standards, Productivity and Innovation Board (SPRING), established in April 2002 as a successor to the Productivity and Standards Board, which had been responsible for managing the productivity and quality performance of Singapore's workforce. Its programmes cover productivity, innovation, and quality, echoing earlier attempts to combine technology assimilation with the transfer of specific modes of labour organisation and industrial relations, as exemplified by experiments with the Japanese practice of quality circles during the 1980s. However, current thrusts of activity are coping with innovation as a matter of skills upgrading. Moreover, quite in accordance with the general orientation of these programmes, the productivity and innovation performance of small- and medium-sized enterprises in the domestic sector is emphasised (Singapore Standards, Productivity and Innovation Board, 2002). The Institute of Technical Education, established as an institution for secondary education and training, governs the domain of workforce training and lifelong learning. Corresponding with the Singaporean strategy of establishing private–public partnerships on an international scale, several of its training programmes are co-organised by multinational enterprises (Institute of Technical Education, 2002). Recent examples have included a Philips Industrial Engineering Programme and an IBM Information Technology Programme for Office Workers. Again, this reflects the fact that skills upgrading is instrumental in Singapore's efforts concerning the formation of an outward-oriented innovation system, supporting its evolution as a knowledge-based economy in the context of globalisation.

4 Conclusion

The Singapore economy is usually portrayed as an example of innovation policies and development strategies that have been outstandingly successful in attracting foreign direct investment for manufacturing and service operations, due to continuous adaptation to the changing conditions of the world economy. Hence, this paper has emphasised that the Singaporean development trajectory has been supported not only by a favourable international environment, but also by responsive policies that have taken advantage of those opportunities which were offered by changes in the dominant techno-economic paradigm. As a result of Singapore's unique situation as a city-state, its experiences are not easily transferable as far as comparison with other newly industrialised economies is concerned. However, the argument of this paper is based on the proposition that the general relevance of the Singaporean case lies in the increasing importance of local agglomerations of high value-added activities which are fuelled by the process of globalisation, for they receive their structural impact from their role as strategic hubs in the networks of multinational enterprises. Therefore, the attraction of foreign direct investment has become a crucial element in governing economic development by technology assimilation. Interactions between government and local and multinational enterprises which characterise the Singaporean innovation system thus demonstrate the paradigmatic character of the Singaporean development experience, which provides lessons for both developed and developing economies in facing the challenge of globalisation.

In particular, this points to the role of multinational enterprises as strategic partners of governments. As illustrated by the case of Singapore, multinational enterprises may introduce novelty into local economic systems, thus contributing to an innovation-driven development process, while government coordinates economic change through the induction of spill-over effects to local enterprises. However, as also in the case of Singapore, an important strategic problem is posed by institution-building in support of these spill-over effects, perceived as a decisive factor in the evolution of local technological capabilities. The Singapore government, while proceeding with these tasks, needs to harmonise its development strategies and innovation policies with the business strategies and long-term objectives of multinational enterprises, a situation which may include conflicting interests in various situations. In particular, the requirement of flexible arrangements may conflict with an integrative development approach that supports the structural cohesion of the local economy. Productivity gaps among small- and medium-sized enterprises in traditional industries provide evidence for that critical aspect. Balancing integration in global innovation networks with the coherence of the local economy is thus a decisive challenge in promoting innovation-driven growth and development.

Notes

1. Globalisation implies that North America, Europe, and East Asia remain the decisive global economic regions which attract flows of trade and investment, while the United States remains the dominant economy in a global technological context (Ebner, 2002: 57n).
2. This is mirrored by universities which exhibit a general trend towards service-oriented interactions with the international business sector (Galli and Teubal, 1997). This tendency echoes the double-sided character of globalisation, namely the parallel internationalisation and local agglomeration of economic activities (Ebner, 2000a: 80n).
3. In assessing these mechanisms, it needs to be reconsidered that firms in latecomer economies tend to reverse the pattern of technological trajectories that is associated with developed economies, ranging from basic production and incremental process change via design, process, and product innovations to competitive R&D (Kim, 1999: 114n).
4. In institutional terms, these development efforts were facilitated by consistent policies and planning procedures, based on the high degree of ideological coherence within government and administration (Vogel, 1991: 76n).
5. This approach had become prominent originally with Taiwan's export processing zones, although the Taiwanese industrial policy approach put more emphasis on small-enterprise networks (Haggard, 1999: 353n).
6. Indeed, the thesis has been put forward that the Singaporean government acts as an entrepreneurial agent in the development process, hence reflecting a policy type that has shaped the East Asian development experience in general (Krause, 1987: 107n).
7. Regarding the theory of entrepreneurship, this pattern is best grasped by the Schumpeterian approach with its distinction of the economic functions of entrepreneurship, namely the carrying out of innovation, and its institutional agents, which may include the government in certain historical situations (Ebner, 2000b: 366n).
8. An important factor in that coordination process is deliberation councils, involving agents from government as well as from the private and public sectors, which play a prominent role over the entire Singapore economy. However, these councils have been widely instituted in all East Asian economies, as they contribute to knowledge flows in concerted policy efforts, thus serving as institutional intermediaries (Aoki, Murdock and Okuno-Fujiwara, 1997: 3n).
9. Indeed, the increase in total factor productivity, that is formally the increase in production output that remains unexplained after increases in the factor inputs have been accounted for in a neoclassical aggregate production function framework, also represented

a common motive for industrial policies in the East Asian newly industrialising economies during that period which was characterised by a more explicit role for innovation and learning (Masuyama, 1997: 5).
10. In empirical terms, these restructuring efforts seem to have promoted an increase in total factor productivity contributions to GDP growth, involving a shift in the output of manufacturing industries, whereas the early growth phase of the Singapore economy during the 1960s and 1970s was primarily driven by factor accumulation, based on the accumulation of capital (Goh and Low, 1996: 9n).
11. Despite national variations, the common characteristics of the East Asian systems of innovation have been portrayed as follows: an expanding education system with an emphasis on tertiary education and engineering, a rapid growth of business in-house R&D, a share of industrial R&D above 50 per cent of gross expenditure on R&D, a rapid development of science and technology infrastructures, and heavy investment in advanced telecommunications. The growth of export-oriented electronics industries then allowed for participation in international technology networks (Freeman, 1996: 178).
12. This assessment holds also with regard to the interaction of multinational enterprises and local firms in general, beyond innovation activities (Mirza, 1986).
13. These data need to be interpreted with the caveat that NSTB statistics define local ownership by a 30 per cent margin, which makes research consortia and joint ventures with a strong qualitative impact of multinational enterprises more difficult to assess, not to mention the subcontractor position of most local enterprises that perform R&D.
14. This is in accordance with the argument that the East Asian development trajectories are generally in need of further impulses regarding R&D and product innovation, as they approach the technological frontier (Hobday, 1995: 200n).
15. Similarly, the notion of the "triple helix", understood as a variant of the systems of innovation approach, accounts for evolutionary interaction and systemic communication between universities, industries, and government. In these relationships, universities play a pivotal role in the production and diffusion of knowledge (Leydesdorff and Etzkowitz, 1997: 155n).
16. This strategic orientation has been of course relevant for all the East Asian newly industrialised economies. It has been maintained that a decisive contribution to the East Asian developmental success, as compared with Latin America, has been provided by sound educational policies which were pursued under the primacy of changing economic and technological requirements, focusing on tertiary education in the area of engineering.

REFERENCES

Amsden, A. and T. Hikino (1993) "Borrowing Technology or Innovating: An Exploration of the Two Paths to Industrial Development", in R. Thomson, ed., *Learning and Technological Change*, New York: St. Martin's Press, pp. 243–266.

Aoki, M., K. Murdock and M. Okuno-Fujiwara (1997) "Beyond the East Asian Miracle: Introducing the Market-enhancing View", in M. Aoki, H.-K. Kim and M. Okunu-Fujiwara, eds., *The Role of Government in East Asian Economic Development: Comparative Institutional Analysis*, Oxford: Clarendon Press, pp. 1–37.

Bell, M. and K. Pavitt (1993) "Accumulating Technological Capability in Developing Countries", in L. H. Summers and S. Shah, eds., *Proceedings of the World Bank Annual Conference on Development Economics 1992*, supplement to the *The World Bank Economic Review*, Washington: The World Bank, pp. 257–281.

Bercuson, K., ed. (1995) *Singapore: A Case Study in Rapid Development*, MSEL International Gov. Pubs. I-IMF 24: 119, Washington, D.C.: International Monetary Fund.

Cantwell, J. (1999) "Innovation as the Principal Source of Growth in the Global Economy", in D. Archibugi, J. Howells and J. Michie, eds., *Innovation Policy in a Global Economy*, Cambridge: Cambridge University Press, pp. 225–241.

Castells, M. (1996) *The Information Age: Economy, Society and Culture, Vol. 1: The Rise of the Network Society*, Cambridge: Blackwell.

Chew, S. B. and R. Chew (1996) "Industrial Relations in Singapore: Past Developments and Future Challenges", in B. K. Kapur, E. T. E. Quah and H. H. Teck, eds., *Development, Trade and the Asia-Pacific*, Singapore: Prentice Hall, pp. 122–140.

de la Mothe, J. and G. Paquet (1996) "Evolution and Inter-creation: The Government–Business–Society Nexus", in J. de la Mothe and G. Paquet, eds., *Evolutionary Economics and the New International Political Economy*, London: Pinter, pp. 9–34.

Department of Statistics (1996a) *Yearbook of Statistics 1995*, Singapore: SNP.
——— (1996b) *Singapore Statistical Highlights 1965–1995*, Singapore: SNP.
——— (2000) *Yearbook of Statistics 2000*, Singapore: SNP.
——— (2001) *Yearbook of Statistics 2001*, Singapore: SNP.

Dosi, G., C. Freeman and S. Fabiani (1994) "The Process of Economic Development: Introducing Some Stylized Facts and Theories on Technologies, Firms and Institutions", *Industrial and Corporate Change* 3(1): 1–45.

Ebner, A. (1999). "Understanding Varieties in the Structure and Performance of National Innovation Systems: The Concept of Economic Style", in J. Groenewegen and J. Vromen, eds., *Institutions and the Evolution of Capitalism: Implications of Evolutionary Economics*, Cheltenham: Edward Elgar, pp. 141–169.

——— (2000a) "Systems of Innovation between Globalisation and Transformation: Policy Implications for the Support of Schumpeterian Entrepreneurship", in A. Kukliński and W. M. Orlowski, eds., *The Knowledge-Based Economy – The Global Challenges of the 21st Century*, Warsaw: State Committee for Scientific Research of the Republic of Poland, pp. 80–99.

——— (2000b) "Schumpeter and the 'Schmollerprogramm': Integrating Theory and History in the Analysis of Economic Development", *Journal of Evolutionary Economics* 10(3): 355–372.

——— (2002) "The Logic of Technological Globalisation: Recent Evidence on R&D in the European Union", in C. Heinrich and H. J. Kujath, eds., *Externe Effekte, Kollektivgüter und Strategien der Regionalisierung*, Münster: Lit, pp. 57–87.

——— (2004) "R&D Cooperation and the Entrepreneurial State: Absorption Capability in the Singaporean System of Innovation", ISEAS Working Paper, Singapore: Institute of Southeast Asian Studies, forthcoming.

Economic Development Board (1996) *Yearbook 1995–1996*, Singapore: EDB.
——— (2000) *Yearbook 1999–2000*, Singapore: EDB.

Economic Planning Committee and Ministry of Trade and Industry (1991) *The Strategic Economic Plan: Towards a Developed Nation*, Singapore: SNP.

Eliasson, G. (2000) "Industrial Policy, Competence Blocs and the Role of Science in Economic Development", *Journal of Evolutionary Economics* 10(1 2): 217–241.

Freeman, C. (1996) "Catching-up and Falling-behind: The Case of Asia and Latin America", in J. de la Mothe and G. Paquet, eds., *Evolutionary Economics and the New International Political Economy*, London: Pinter, pp. 160–179.

Galli, R. and M. Teubal (1997) "Paradigmatic Shifts in National Innovation Systems", in C. Edquist, ed., *Systems of Innovation: Technologies, Organizations and Institutions*, London: Pinter, pp. 342–370.

Goh, K. S. and L. Low (1996) "Beyond 'Miracles' and Total Factor Productivity: The Singapore Experience", *ASEAN Economic Bulletin* 13(1): 1–13.

Haggard, S. (1999) "An External View of Singapore's Developed Status", in L. Low, ed., *Singapore: Towards A Developed Status*, Oxford: Oxford University Press, pp. 345–375.

Ho, K. C. (1993) "Industrial Restructuring and the Dynamics of City-State Adjustments", *Environment and Planning A* 25(1): 47–62.

Hobday, M. (1995) *Innovation in East Asia: The Challenge to Japan*, Aldershot: Elgar.

Huff, W. G. (1994) *The Economic Growth of Singapore: Trade and Development in the Twentieth Century*, Cambridge: Cambridge University Press.

Institute of Technical Education (2002) *Breakthrough Changes: Annual Report 2001/02*, Singapore: ITE.

Islam, I. and A. Chowdhury (1997) *Asia-Pacific Economies: A Survey*, London: Routledge.

Kim, L. (1999) "Building Technological Capability for Industrialization: Analytical Frameworks and Korea's Experience", *Industrial and Corporate Change* 8(1): 111–136.

Krause, L. B. (1987) "The Government as an Entrepreneur", in L. B. Krause, K. A. Tee and L. Yuan, eds., *The Singapore Economy Reconsidered*, Singapore: Institute of Southeast Asian Studies (ISEAS), pp. 107–127.

Leydesdorff, L. and H. Etzkowitz (1997) "A Triple Helix of University–Industry–Government Relations", in H. Etzkowitz and L. Leydesdorff, eds., *Universities and the Global Knowledge Economy: A Triple Helix of University–Industry–Government Relations*, London: Pinter, pp. 155–162.

Low, L. (1998) *The Political Economy of a City-State: Government-made Singapore*, Singapore: Oxford University Press.

Low, L., M. H. Toh, T. W. Soon, K. Y. Tan and H. Hughes (1993) *Challenge and Response: Thirty Years of the Economic Development Board*, Singapore: Times Academic Press.

Mani, S. (2000) "Policy Instruments for Stimulating R&D in the Enterprise Sector: The Contrasting Experiences of Two MNC Dominated Economies from Southeast Asia", UNU-INTECH Discussion Paper 2000-9, Maastricht: UNU-INTECH.

Masuyama, S. (1997) "The Evolving Nature of Industrial Policy in East Asia", in S. Masuyama, D. Vandenbrink and C. S. Yue, eds., *Industrial Policies in East*

Asia, Tokyo: Nomura Research Institute; and Singapore: Institute of Southeast Asian Studies (ISEAS), pp. 3–18.

Ministry of Trade and Industry (1986) *The Singapore Economy: New Directions*, Singapore: SNP.

—— (1997) *Economic Survey of Singapore 1996*, Singapore: SNP.

—— (2001) *Economic Survey of Singapore 2000*, Singapore: SNP.

Mirza, H. (1986) *Multinationals and the Growth of the Singapore Economy*, London: Croom Helm.

Nanyang Technological University (1996) *Research in NTU*, Singapore: NTU.

National Computer Board (1992) *A Vision of an Intelligent Island: IT 2000 Report*, Singapore: SNP.

National Science and Technology Board (1991) *National Technology Plan 1991: Window of Opportunities*, Singapore: SNP.

—— (1999) *Singapore Technology Infrastructure*, Singapore: NSTB.

—— (2000) *National Survey of R&D in Singapore 1999*, Singapore: NSTB.

National University of Singapore (1996) *Technology and Expertise Directory*, Singapore: NUS.

Porter, M. (1990) *The Competitive Advantage of Nations*, New York: Free Press.

Rodan, G. (1989) *The Political Economy of Singapore's Industrialization: National State and International Capital*, London: Macmillan.

Saxenian, A. L. (1994) *Regional Advantage: Culture and Competition in Silicon Valley and Route 128*, Cambridge, Mass.: Harvard University Press.

Schein, E. H. (1996) *Strategic Pragmatism. The Culture of Singapore's Economic Development Board*, Singapore: Toppan and MIT Press.

Singapore Standards, Productivity and Innovation Board (2002) *Annual Report 2001/02*, Singapore: SPRING.

Soon, T.-W. and C. S. Tang (1993) *Singapore: Public Policy and Economic Development*, Washington: World Bank.

Vogel, E. F. (1991) *The Four Little Dragons: The Spread of Industrialization in East Asia*, Cambridge, Mass.: Harvard University Press.

Wong, P. K. (1995) "National Innovation System: The Case of Singapore", ASEAN-ROK Cooperation Project Working Paper, Seoul: Science and Technology Policy Institute.

—— (1999) "University–Industry Technological Collaboration in Singapore: Emerging Patterns and Industry Concerns", *International Journal of Technology Management* 18(3/4): 270–283.

Wong, P. K. and C. Y. Ng (1997) "Singapore's Industrial Policy to the Year 2000", in S. Masuyama, D. Vandenbrink and C. S. Yue, eds., *Industrial Policies in East Asia*, Tokyo: Nomura Research Institute; and Singapore: Institute of Southeast Asian Studies (ISEAS), pp. 121–141.

World Bank (2000) *World Development Report 1999/2000: Entering the 21st Century*, Oxford: Oxford University Press.

Yun, H. A. (1998) "Innovative Milieu and Cooperation Networks: State Initiatives and Partnership for Restructuring in Singapore", in H.-J. Braczyk, P. Cooke and M. Heidenreich, eds., *Regional Innovation Systems: The Role of Governance in a Globalized World*, London: UCL Press, pp. 379–412.

4

Evolution of the civil aircraft manufacturing system of innovation: A case study in Brazil

Rosane Argou Marques

1 Introduction

The literature on evolutionary economics considers the network of actors interacting in order to promote innovation, which is under a certain institutional infrastructure, at the heart of dynamic economies. The system of innovation approach has made important contributions to the understanding of this issue. These contributions are generally concerned with the relationship between economic growth and technological development and basically focus on two modes of analysis: (i) geographical differences, mainly between countries and regions (national or regional systems), regarding the type of exported products, investments in R&D, investments in education and training, science and technology capabilities, industrial structure, patents, etc.; and (ii) technological diffusion and development within industrial networks (technological or sectoral systems).

Systems of innovation are understood to be the networks of government and non-government agencies, science and technology institutes, educational organisations, firms, and other organisations. The country's macro-economic and industrial policies, international regulations, market governance, and socio-cultural institutions influence the network dynamism and trajectory. The interaction between the former and the latter has influenced knowledge accumulation and learning processes in firms (Cooke, Uranga and Etxebarria, 1997; Freeman, 1987; Nelson and Rosenberg, 1993). To understand this issue, many authors have focused on distinct but inter-related areas of systems of innovation. Some of these areas are related to technological, sectoral, national, regional, financial, and political systems of innovation, etc. This chapter focuses on examining the evolution of a sector, specifically its dynamics and transformation over time with regard to technologies and relationships between actors. Although much research has been conducted on this theme, the structure and changes in the relationships among actors in the system are not fully understood (Malerba, 2002). Therefore, this chapter aims to examine this issue at a sectoral level.

Specifically, the main aim of this chapter is to address how the Brazilian civil aircraft manufacturing system of innovation (BASI) has developed since the privatisation of the aircraft producer Embraer in 1994. The development of this innovation system is concerned with the examination of the key actors and the characteristics of their relationships with each other. Specifically, it explores the relationships between the aircraft designer and producer (Empresa Brasileira de Aeronáutica S. A. Embraer), tier one foreign suppliers, tier two local suppliers, and the Aerospace Technical Centre (CTA). Thus, changes in the relationships between these actors in the system are described. It is expected that it will be found that the relationships between the actors have evolved so that first tier suppliers in Embraer's product development activities have assumed an increasing role since 1994.

Compared with other countries with similar industrialisation policies, Brazil has built up a large scientific and technological capability in the aircraft manufacturing sector (Goldstein, 2001), whose sales corresponded approximately to 1.1 per cent of the Brazilian total GNP in 2000 (Arnt, 2001). The aeronautical science and technology infrastructure and the aircraft manufacturing sector were established under government import substitution industrialisation policies (1950–1989). Embraer was the largest Brazilian exporter in 2000 (Arnt, 2001), with sales of approximately US$2.7 billion, making it the world's fourth largest manufacturer of civil aircraft (Fong, 2001). The Ministry of Aeronautics founded the company in 1969 and privatised it in 1994.

The aircraft producer, founded in 1969 as a spin-off from the Brazilian CTA,[1] has developed capabilities through internal investments in knowledge accumulation and acquisition, and interactions with firms and technological institutes (Frischtak, 1992, 1994). The majority of its aircraft systems, structural parts, components, and sub-systems are imported, illustrating that Brazil lacks local supply chain capabilities. This deficiency, together with the growth of the aircraft producer, is one of the main problems discussed by Brazilian researchers and policy makers.

The present situation in the Brazilian aircraft manufacturing sector has focused concern on the role of government science and technology

(S&T) policies and the ability of Embraer to strengthen the local supply chain and reduce imports. Therefore, research on this industry has been directed towards the sector's competitiveness (Coutinho and Ferraz, 1993; Dagnino, 1993) and the impact of liberalisation policies and consequent structural reform in the aeronautical productive arrangement (Bernardes, 2000a,b). The ways the aircraft producer has acquired technological capabilities since its foundation in 1969 are also being researched (Frischtak, 1992, 1994).

In order to understand the structure and changes in the BASI, data were gathered in the form of semi-structured interviews conducted with three non-governmental organisations, two universities, four aeronautical technological institutes, Embraer, and 20 suppliers in 2001 and 2002. The results were compared with existing literature and other non-interview data sources, which included academic and industry literature, documentation, archival records, direct observation, and physical artefacts (Ellis et al., 2000).

The analytical framework is based on the systems of innovation literature, which is explained in section 2. Two levels of disaggregation (production system and knowledge system) are considered in order to examine the trends in the relationships between Embraer, local suppliers, foreign suppliers, and technological centres. The case study of the BASI is then examined taking into consideration its technological and economic performance (section 3) and the main changes in relationships (section 4). Section 5 discusses the main changes in the production and knowledge systems, which have raised questions about improving the capabilities of Brazilian suppliers to allow them to increase their participation in Brazilian civil aircraft manufacturing.

2 The literature on systems of innovation: Some features and questions

The theory that technical change is not an isolated process emerged as an attempt to explain innovative behaviour and the consequent technological capability accumulation and evolution in firms (Freeman, 1987). Technical change is therefore a consequence of the capability of firms to manage and generate innovation as well as acquire and diffuse technological knowledge. Its development is a process that requires the involvement of other firms, universities, and government institutions among other organisations. Government policies may also play an important role in regulating and coordinating the pace (quantity) and nature (quality) of the development of technological capabilities according the country, region, and/or sector-specific endowments and characteristics (De Ferranti and Perry, 2002).

The literature on systems of innovation (SI) has made important contributions to the understanding of innovation as an interactive learning process. Innovation is considered as an "interactive process, which involves the creation of qualitatively different new things and new knowledge" (Lundvall, 1992: 46). The literature generally focuses on product innovation and interaction between user-producers in organised markets and great importance is given to the flow of qualitative information about the production and utilisation of the new product (Lundvall, 1992). There are basically two modes of analysis focusing on: (i) geographical differences, mainly between countries and regions (national or regional systems), regarding the type of exported products, investments in R&D, investments in education and training, science and technology capabilities, industrial structure, and patents, etc.; and (ii) technological diffusion and development within industrial networks (technological or sectoral systems).

The common features in SI definition are that it involves a network of government and non-government agencies, science and technology institutes, firms, and educational organisations, etc. The network dynamics and trajectory are influenced by the country's macro-economic and industrial policies, international regulations, market governance, and socio-cultural institutions, and interactions between these influence the innovation process (Freeman, 1987; Lundvall, 1992; Malerba, 2002; Nelson and Rosenberg, 1993). The flows of qualitative information or knowledge are mainly associated with the flows of tangible (goods and services) "capital" (Lundvall, 1992). At least one question arises from the literature: Are developing countries also organised in "systems of innovation" or they are "learning systems" as referred by Viotti (2002)?

Viotti (2002) considers that developing countries are adopters of technological knowledge from developed countries, though their firms may develop incremental innovations according to their capabilities to do so. However, these firms do not develop innovations in the sense defined by Lundvall (1992) and Schumpeter (Malerba, 2002). Viotti compares the cases of South Korea and Brazil and concludes that the former is an active learning system and the latter a passive learning system. An active learning system is characterised by the capability to improve and adapt technological capabilities, while a passive learning system is characterised by the capability to adopt new technologies.

The conclusion that the Brazilian system of innovation, like that of other Latin American countries, is not innovative is supported by other research. Katz (2000, 2001), Bernardes (2000b), and Cassiolato and Lastres (1999) examined the technological behaviour of national and foreign firms and the influence on this behaviour of macro-economic policies and industrialisation strategies defined by governments. Their common conclusion is that there are weak linkages between firms and S&T infrastructure in the national and local systems of innovation in Latin American countries and a lack of long-term industrial policies for technological development. Therefore, they conclude that local firms have lacked technological capabilities to succeed in competing in foreign markets.

Cassiolato and Lastres (1999) and Dahlman and Frischtak (1990) reported that the structure of the Brazilian national system of innovation is heavily influenced by government policies. Specifically, the structural changes in the system resulting from import substitution industrialisation policies have maintained and strengthened the role of imported technologies and subsidiaries of foreign firms in indigenous technological development. Dahlman and Frischtak (1990) observed that by 1960 more than 50 per cent of the total goods manufactured in Brazil were produced by subsidiaries of foreign corporations. They also concluded that the overall S&T infrastructure was developed by the government for improving technological capabilities and developing a local supply chain to support the foreign subsidiaries' production facilities and the national state-owned firms. Apparently this picture has not changed and Quadros et al. (2001) concluded that local foreign-controlled firms accounted for the largest share of private R&D activity, which is concentrated in the adaptation of products and processes originating from abroad. This conclusion is supported by other Brazilian researchers (Costa and Queiroz, 2002) who mention a "moderate" improvement in Brazilian firms' technological capability after the Brazilian government shifted from import substitution industrialisation policies to liberalization policies, although the improvement is more effectively accomplished by local subsidiaries of foreign firms.

However, there are some successful cases of the development of technological capabilities through explicit government policies in Brazilian industry, as in the case of Petrobrás (exploration of offshore petroleum) and Embraer (design and production of short-haul jets) (De Ferranti and Perry, 2002). There are some questions regarding these two cases: Were the government efforts successful in terms of the increasing international competitiveness of the firms? What is the participation of local Brazilian suppliers in the sector? Are there weak linkages between firms and S&T infrastructure as observed through research of the national system of innovation?

In order to examine these questions, a sectoral approach to the system of innovation is adopted. Specifically, two levels of sectoral disaggregation in terms of type of flow are examined: the production system and the knowledge system. The next sub-section examines their definitions as well as how the two systems interact in the civil aircraft manufacturing industry in order to analyse the case of Embraer. This interaction is considered by Ruffles (1992) as an important aspect of technological dynamism in this sector.

2.1 Production and knowledge systems in civil aircraft manufacturing

The main aim of this section is to contribute to the discussion on the interaction between the knowledge system (KS) and the production system (PS) in civil aircraft manufacturing. Bell and Albu (1999) define the knowledge system as bounded by inter-organisational "stocks and flows of knowledge", while the production system is bounded by inter-firm "product designs, materials, machines, labour inputs, and transaction linkages involved in production of goods to a given specification" (pp. 1722–1723). Therefore, the KS focuses on the flow of knowledge irrespective of its connection to the flow of goods, with the main objective of developing and implementing technological change in firms (Gelsing, 1992). In this sense, the KS is an important element of a sectoral system of innovation. The inter-organisational stocks and flows of knowledge contribute to technological learning processes in firms, which may stimulate sectoral technological capabilities development.

The interaction between the KS and PS is examined taking into consideration three elements: (i) the structure of the civil aircraft production chain; (ii) the type of technological knowledge that supports products, processes, operational procedures, and facilities development; and (iii) the instruments utilised for promoting the interaction. The general structure of the supply chain for civil aircraft production is hierarchically divided into five tiers (based on Hwang, 2000; Ruffles, 1992; and interviews):

(i) Aircraft designer and producer
(ii) Airframe structures and major systems, such as propulsion system, environmental system, hydraulic system, electric system and avionics system
(iii) Navigation, communication, surveillance, and flight management systems, engine systems, starting systems and electrical power structures, interior cabin systems and components, air conditioning and pressurization system, oxygen system, electronic/electrical components, landing gear systems, and fuel system, among others
(iv) Avionics parts and components, engine parts and components, fuselage parts and components, electric/electronic parts and components, among others
(v) Other suppliers

S&T institutes, firms in the production system, and other organisations form the knowledge system whose knowledge development may effect changes in the product, the processes, operational procedures, and facilities technologies. There are two types of technological knowledge flows:

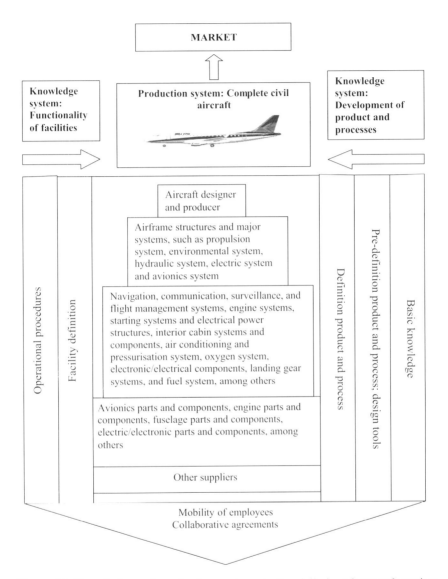

Figure 4.1 Knowledge and production systems in the civil aircraft manufacturing system of innovation.
Source: Based on Hwang (2000), Ruffles (1992), and interviews.

one supports the definition of product and processes (right side of fig. 4.1) and the other supports organisational procedures and facilities improvements (left side of fig. 4.1). These two systems are closely related at all levels in the hierarchical supply chain and are connected through formal and informal collaborative agreements and the mobility of employees.

Basic knowledge is concerned with the production of new scientific knowledge that influences advances in product and processes technologies. These advances and the perception of market needs guide the pre-definition of new products and processes or their general concept development, and the design tools. Thus, detailed definition of product and processes is achieved within a great variety of technological possibilities. At the same time, decisions are made concerning operational and facilities alternatives. Finally, the product and processes are entered into the production phase. The production phase can also result in improvements in the product and process detailed definition, and in operational choice. These phases of knowledge development are not linear and the organisational processes for their integration define the dynamic of interactions.

Interaction between the knowledge system and the production system depends on the complexity of the learning processes and knowledge accumulation, considered to be a result of the novelty of embodied technologies and complexity of component integration in developing the final product. Therefore, tiers one and two in the supply chain may be the most integrated with the civil aircraft producer and with the knowledge system. The aircraft producer and the tier one suppliers are probably those that transmit downwards the knowledge of new product and process technologies while they are highly dependent on each other for future aircraft development.

The civil aircraft manufacturing system of innovation development is the result of integration between the knowledge system and the production system explained in figure 4.1. There are at least two main ways to promote this integration: (i) by incentives for training and qualification of engineers and technicians who will move within the supply chain and between it and S&T institutions; and (ii) through collaborative agreements between actors in the knowledge system. This chapter focuses on examining the collaborative agreements in the civil aircraft manufacturing system of innovation.

According to Ruffles (1992), the main collaborative agreements in the civil aircraft manufacturing system of innovation are: joint ventures, risk-sharing, preferred supplier arrangement, and technology collaboration. The term "joint venture" relates to the foundation of a joint company for developing a specific programme. The partners can be competitor firms or not. Technological transfer is limited to that necessary to perform the programme. A "risk-sharing agreement" is normally between firms from different tiers in the supply chain. Supply firms take a share in the costs of the programme development and in its revenues as the proportion of their investment. Thus, technological knowledge improvement depends on each party's own development. This partnership lasts for the life of the product and it is unlikely that re-sourcing occurs. A "preferred supplier arrangement" presupposes less commitment between the parties compared with joint venture and risk-sharing agreements. It is based on long-term understanding between the customer and supplier and is subject to satisfactory performance, which is defined by the "request for proposals" and formal or informal contracts. In general, the customer and supplier work together to achieve lower prices, higher quality, and greater efficiency. According to Ruffles (1992: 7), this arrangement normally "applies at lower levels of the production chain and is applicable to the acquisition of both product and fundamental technology". He states that "typically, eighty per cent of the product cost is locked in at the design phase and the aim of close partnership with hardware suppliers is to utilise their technical expertise during product definition in order to achieve the lowest cost solution" (Ruffles, 1992: 7). "Technology collaboration" describes cooperation for pre-competitive technological development with the objective of generating knowledge in key technologies and improving the basic knowledge in the sector. This type of collaboration is joint funded by governments and is focused on long-term development. These collaborative agreements, as well as the other forms of collaboration and the mobility of engineers within the industry, are important for integrating the knowledge and production systems (fig. 4.1).

These observations explain the dynamics of the civil aircraft system of innovation regarding the integration of the knowledge and production systems for developing new technologies. But there is little understanding about the dynamics of the integration between the knowledge and production systems when a factor other than "new technologies" affects the system. It seems that this is the case with the Brazilian civil aircraft system of innovation (BASI), where major changes concerned changing the ownership of the main aircraft producer, the role of technological institutions, technological complexity, and structure of the supply chain, etc. Therefore, some important characteristics of the BASI are examined in section 4. First, however, the distinct phases of the technological development of Brazilian civil aircraft manufacturing are briefly described in section 3.

3 Changing features of Brazilian civil aircraft manufacturing

The main objective of this section is to examine how Brazilian civil aircraft manufacturing has been developed since the foundation of Embraer. Using an historical approach, this section will give an overview of the steps followed by Embraer to become the world's fourth largest aircraft producer (Fong, 2001). In order to do so, some features of the Brazilian aircraft industry in the period before the foundation of Embraer are briefly explained in section 3.1. Section 3.2 then examines the main characteristics of its three phases of development: (i) start-up (1969–1978); (ii) entry into the international market (1979–1994); and (iii) post-privatisation (1995–2002). Finally, the general market characteristics are described in section 3.3.

3.1 The Brazilian aircraft industry in the period before 1969

The period before 1969 can be examined by considering the three important periods of development of the Brazilian aircraft industry. The first period was during the decades of 1910–1930. According to Dagnino and Proença (1989), the first step towards developing an aircraft industry in Brazil was taken in the decade following 1910 when Santos Dumont, a Brazilian industrialist, developed the first heavier-than-air flying machine. Although there was some investment during this period, Brazil lacked the engineering and therefore technological capabilities, as well as the government support, that the industry required. Therefore such efforts as there were did not result in any significant development of the Brazilian aircraft industry.

The second period was during the Second World War (1935–1945). Brazil, as an ally of the United States, produced fighter aircraft. During this time, the American Air Force trained Brazilian pilots and aeronautical engineers to help produce approximately one aircraft per day for the United States (Dagnino and Proença, 1989). Although more effort was expended than in the first phase, the Brazilian aircraft industry improved little, due to the continuing lack of technological and engineering capability to design and produce an entirely Brazilian aircraft. Aircraft production was restricted to the assembly of foreign military aircraft models and small aircraft for agricultural and training use.

Therefore, the Brazilian Defence Ministry decided to create an aeronautical institute to produce highly qualified engineers to support the infant aircraft industry. The Technological Institute of Aeronautics (ITA) was founded at the end of the 1940s, which marks the beginning of the third period. The ITA was set up with the support of MIT and NASA, and by 1970 had produced approximately 2,000 engineers, who were mostly contracted to work in other sectors due to lack of aircraft companies (Bernardes, 2000a; Dagnino, 1993; Dagnino and Proença, 1989).

During the 1950s, the Brazilian army was aware that it was also necessary to create a research centre to apply aeronautical engineering knowledge to the development of a "Brazilian aircraft" adapted to the

Brazilian territorial endowments and characteristics. They founded the Aerospace Technical Centre (CTA), which absorbed the ITA, and also developed other institutes for aeronautical research. The main research project at the CTA was to design and produce a 19-seater aircraft. The Brazilian Ministry of Aeronautics contracted an entire research group from Germany to work with Brazilian engineers from ITA to develop this aircraft. The first prototype flew in 1959, but further improvements were necessary. In 1969, the project finished with the production of the first aircraft, the Bandeirante. In 1969, the group of researchers then set up Embraer, the first state-owned aircraft producer in Brazil, with the support of the Brazilian Ministry of Aeronautics (Bernardes, 2000a; Dagnino, 1993; Dagnino and Proença, 1989).

3.2 The Brazilian aircraft industry from 1969 to 2002

The "start-up phase" began in 1969 when the Brazilian Ministry of Aeronautics founded the Empresa Brasileira de Aeronáutica S. A. Embraer. The aircraft producer was created as a spin-off from the CTA, with the objective of supplying the Brazilian Air Force with parts, components, and training and fighter aircraft (Bernardes, 2000a; Coutinho and Ferraz, 1993; Dagnino and Proença, 1989). According to Frischtak (1994: 602), "although the production of airplanes in Brazil dates back to 1910, when the first monoplane was built in the country, the development of the Brazilian passenger aircraft industry can be equated with the development of Embraer". The main civil aircraft produced during the 1970s was a 19-seater light twin-engine turbo propeller plane (Bernardes, 2000a; Frischtak, 1992).

The Ministry of Defence was the main buyer and also gave strong tax incentives and subsidies to Embraer to develop production and technological capabilities to manufacture the 19-seater aircraft (Bernardes, 2000a; Coutinho and Ferraz, 1993; Dagnino and Proença, 1989). These incentives were oriented towards financing (through subsidies and tax exemption), marketing (through procurement and protectionism), and developing technologically (through special decrees for technology transfer and supporting research). In the first 10 years of existence, the aircraft produced were mainly sold within Brazil.

According to Dagnino and Proença (1989), although the Ministry of Defence had heavily invested in the creation of a national aircraft supply chain (aircraft assembler and suppliers), approximately 68 per cent of parts, components, and sub-systems of aircraft produced were imported. Nevertheless, some suppliers, such as Aeromot (located in Porto Alegre, State of Rio Grande do Sul) and Neiva (Botucatú, State of São Paulo), progressed from producing parts and components to producing small aircraft (1–10 seats). This development was possible due to the Ministry of Defence special programmes for "nationalisation" of aero parts (systems, structural parts, and other components). However, according to interviews at the Institute for Development and Coordination of the Aerospace Industry (IFI), low production scale, high quality, and the high development costs of aero parts influenced the concentration of production at Embraer. Few local supplier firms developed the capacity to produce parts and components, and relied heavily on technological transfer from the CTA through IFI consultancy. Therefore, most aero parts were still being imported at the end of the 1970s.

The "international market-seeking phase" corresponds to an increase in exports of small-bodied aircraft (10–30 seats) which occurred after the American market was deregulated in 1978 (Coutinho and Ferraz, 1993). Therefore, the second phase corresponds to the period when the market changed from being national to being foreign. A 30-seater advanced twin-engine turbo propeller aircraft was the main commercial product during this phase. According to Frischtak (1992: 13), "At end 1990, [the 30-seater aircraft] market share in the 20–45 seat category was 25 per cent worldwide, just slightly below that of its major competitor (the SAAB SF340). In the US market, [it] had the dominant position in that year in terms of the total number of aircraft in service, again for the 20–45 seat category."

The launching of an eight-seater twin-engine turbo propeller pressurised aircraft in 1979 is the starting point of this phase (Bernardes, 2000a; Frischtak, 1994). This aircraft was the first entirely undertaken by Embraer. Its main market was not the Brazilian Air Force but large American corporations: it was a business aircraft designed to service the American market. The aircraft producer undertook product development, financed it, and designed and manufactured the pressurised system (one of the main innovations in this model) (Frischtak, 1994). It was the first aircraft developed using the concept of communality or "family".[2] The second in the family was a 30-seater aircraft, which was launched in 1981 to supply the United States and Latin American markets (Frischtak, 1992).

The "nationalisation" of the aero parts programme that began in the start-up phase ceased almost completely during this phase. The phase of "denationalisation" began as the aircraft producer focused on technological development aimed at international market standards. Another important characteristic of this phase was the increased reliance on imported systems, components, and sub-systems. The new market demanded many improvements in digital technology, new materials, and sophisticated software among other technological developments which the local suppliers could not provide. Local suppliers lacked government incentives and economies of scale for this technological upgrading. Also, the Ministry of Defence had reduced the IFI budget, which significantly decreased IFI support for the technological development of local suppliers.

The recession in the international civil aircraft market and the fact that the Brazilian government decreased subsidies and ordered fewer aircraft were the main factors affecting the financial crisis of Embraer at the beginning of the 1990s (Bernardes, 2000a,b) and the company was privatised in 1994. Many small- and medium-sized local Brazilian suppliers exited the market due to the economic recession of 1990–1994.

The most important products in the "post-privatisation period" have been the ERJ 145 and ERJ 170 jetliners (Bernardes, 2000a,b). The ERJ 145 jetliner has the basic platform of the 30-seater advanced turbo propeller aircraft but incorporates new technologies in avionics, propulsion, and aerodynamics, and was launched in 1995 (Bernardes, 2000a). The ERJ 170 jetliner was first "rolled-out" in November 2001, with the first flight in 2002. Both aircraft were developed within the concept of "family" or "commonality".

Summarising, the Brazilian government has supported civil aircraft manufacturing in all stages of its development basically through (Bernardes, 2000a; Green, 1987): (i) research and development policies; (ii) joint government–private ownership; (iii) protection of national markets; and (iv) export development policies. However, at least one important question is: Were the government efforts successful in terms of increasing the international competitiveness of civil aircraft manufacturing?

3.3 *General characteristics of the Brazilian civil aircraft market*

According to Bernardes (2000a), Dagnino (1993), and Donângelo, Coelho and Ichimura (2000), the Brazilian international civil aircraft market is mainly the United States and Europe, although there is investment to increase participation in China and Asia. During the 1970s and 1980s, the main market, however, was Brazil, whose imports were restricted by the "Law of Similars" (Chapter III, Section V of Decree-Law No. 37, as implemented by Decree No. 61,574 of October 20, 1967) which was part of the import substitution industrialisation policies (Green, 1987). Embraer was then granted the monopoly for production and commercialisation of aircraft turboprops with more than eight seats. Piper was the only foreign Embraer competitor selling in the Brazilian market due to a licence agreement signed between them before the Law of Similars was implemented. During this period, the main competitors of Embraer in the United States and Europe were: De Havilland, Cessna, Fairchild, Piper, Saab, BAe, Dornier, Fokker, and Canadair.

By the end of the 1980s and during the 1990s, some civil aircraft manufacturers exited the civil aircraft market, such as BAe, Cessna, Saab, and Fokker, while others merged or were acquired, such as Fairchild-Dornier and Bombardier-Canadair. Fairchild-Dornier filed for bankruptcy in 2002. Since the 1990s, the main international competitor of Embraer has been Bombardier-Canadair, otherwise called Bombardier Aerospace. Bombardier is the third largest world producer of regional jets (Padgett, 2003),[3] while Embraer is the world's fourth largest with a 45 per cent share of the regional jet market in 2000 (Fong, 2001). The fierce competition between the two companies led them to complain to the World Trade Organization about unfair subsidies given by the Brazilian and Canadian governments. Bombardier complains that Embraer's jets are less technologically advanced than their jets and are doing well in the market due to lower labour costs, the cheap Brazilian currency, and Brazilian government subsidies. Embraer complains that Bombardier's jets are subsidised by the Canadian government's low loan rates. The complaints to the WTO started in approximately 1998 and are still continuing (Padgett, 2003).

The main civil aircraft models manufactured by Embraer are shown in table 4.1. Embraer's market segment ranges from small turboprops seating 8–30 passengers, developed during the 1970s and 1980s, to medium-sized jets seating 35–108 passengers. These planes fly specifically short hauls or regional routes, mainly linking hub routes and small airports. $A_s{}^*V_c$ is the performance indicator that represents the number of seats (A_s) multiplied by the speed (V_c). This is considered by Mowery and Rosenberg (1981, 1985) to be an important indicator of performance development. Table 4.1 shows that there has been a substantial increase in Embraer aircraft performance since the launch of the ERJ 145 jetliner.

At the end of the 1980s, the Brazilian government started to open up the national market and changed the macro policies from import substitution industrialisation to liberalisation. It substantially reduced the subsidies given to Embraer and there were no longer import restrictions on small regional aircraft as was the case in the previous period. Embraer, which was controlled by the Ministry of Aeronautics, was sold to private companies in 1994. The voting shares are owned by Bozano Simonsen (Brazil's biggest investment company), 20 per cent; PREVI (Brazilian pension fund), 20 per cent; SISTEL (Brazilian pension fund), 20 per cent; a French group (composed of EADS, Dassault Aviation, Snecma, and Thomson-CSF),[4] 20 per cent; the Brazilian government, 1 per cent; and Bovespa (the Brazilian stock exchange), 19 per cent. The company also launched shares to be negotiated in the New York Stock Exchange (NYSE) in 2000. The preference shares are owned by Bozano Simonsen, 20 per cent; PREVI, 20 per cent; SISTEL, 13 per cent; NYSE, 29 per

Table 4.1 The evolution of aircraft models produced in Brazil

Year (first plane flew)	Model	Seats	Altitude (feet)	Speed (km/h)	Characteristics	$(A_s * V_c)$
1972	EMB 110 Bandeirante	19	22,500	413	Light twin turboprop	7,847
1979	EMB 121 Xingú	8	26,000	450	Twin turboprop pressurised	3,600
1983	EMB 120 Brasília	30	30,000	555	Turbo propeller pressurised	16,650
1995	ERJ 145	50	37,000	833	Twin turbofan (Jet)	41,650
1995	ERJ 140	44	37,000	833	Twin turbofan (Jet)	36,652
1998	ERJ 135	37	37,000	833	Twin turbofan (Jet)	30,821
2002[a]	ERJ 170	70	37,000	870	Jet	60,900
2003[a]	ERJ 190–100	98	37,000	870	Jet	85,260
2004[a]	ERJ 190–200	108	37,000	870	Jet	93,960

Source: Websites http://www.embraer.com and http://www.airliners.net/.
Note: Information gathered in January 2001.
a. Year planned for the first flight.

cent; and others, 18 per cent.[5] The Brazilian Ministry of Defence has 1 per cent of shares, which are called "golden shares" because its rights of voting are restricted mainly to defence matters. According to Padgett (2003), Embraer's privatisation in 1994 heralded Brazil's new push to be a global economic player in the civil aircraft market.

Nowadays, Brazilian aircraft manufacturing industry sales are mainly to foreign markets, accounting for approximately 72.5 per cent of turnover in 2000. The United States was responsible for approximately 60 per cent of these exports in 1999 according to Bernardes (2000a,b, 2001). AIAB (2001) observed an increase in total exports from US$0.70 billion in 1997 to US$2.50 billion in 2000 (table 4.2). As a consequence, the aircraft industry has increased its contribution to Brazilian GDP, measured by total turnover divided by total GDP, which jumped from 0.29 per cent in 1997 to 1.06 per cent in 2000 (table 4.2). Embraer represents approximately 80 per cent of the data shown in table 4.2.

Table 4.3 shows the economic performance of Embraer from 1990 to 2002, which has been positive in terms of net profit since 1997. The decrease in turnover from 1990 to 1994, as shown in table 4.3, was caused by the international crisis in the aircraft industry and a reduction in pro-

Table 4.2 Economic performance of the Brazilian aircraft manufacturing sector

Economic indicators	1996	1997	2000
Turnover (US$ billion)	0.60	0.80	3.40
Turnover/Brazilian GDP (%)	0.22	0.29	1.06
Exports (US$ billion)	0.20	0.70	2.50
Employees	6,500	8,000	13,500

Sources: Bernardes (2000a), AIAB (2001).

curement by the Brazilian government. The suggested causes for the increase since 1997 are: (i) the international success of the ERJ 145 jetliner; and (ii) expansion in the international commercial short-haul market. Actually, the international crisis in this market, which was accentuated by the September 2001 attacks in the United States, is having a negative impact on the sector worldwide, with a possible reduction of 20 per cent in Embraer's planned sales (Bernardes, 2001; and interviews). According

Table 4.3 Embraer economic indicators, 1990–2002

Year	Turnover (US$ billion)	Net profit (US$ billion)	Exports (US$ billion)	Brazilian market[b] (US$ billion)	Employees
1990	0.58	−0.26	0.21	0.37	–[c]
1991	0.40	−0.24	0.12	0.28	–
1992	0.33	−0.25	0.10	0.23	–
1993	0.26	−0.11	0.10	0.16	–
1994	0.17	−0.31	0.06	0.11	6,087
1995	0.29	−0.25	0.11	0.18	4,319
1996	0.38	−0.12	0.13	0.25	3,849
1997	0.83	−0.03	0.69	0.14	4,494
1998	1.58[a]	0.07[a]	1.17	0.41	6,737
1999	1.77[a]	0.21[a]	1.69	0.08	8,302
2000	2.75[a]	0.33[a]	2.70	0.05	10,334
2001	2.97[a]	0.57[a]	2.89	0.08	9,218
2002	3.32[a]	0.62[a]	2.39	0.93	10,097

Sources: Bernardes (2000a), Embraer Administration Report (2000, 2001, 2002) downloaded on 8 July 2003 from http://www.embraer.com.br.
a. Values converted from Brazilian reais to the US dollar using the exchange rate of the year 1999, R$1.9 per US$1.00. Therefore, the values are indicative.
b. The participation of the Brazilian market is calculated by subtracting exports from turnover.
c. This indicates that it was not possible to get the information by the time this chapter was being finished.

to Padgett (2003), Embraer stock has fallen from a high of US$30 before the September 2001 attacks to about US$15, while Bombardier shares are stuck at around US$3, down from about US$18.

Although Embraer has substantially improved its position in the international market, Brazilian researchers and policy makers are concerned about the rising component imports and the development of the national supply chain. Many changes have occurred in the structure of the civil aircraft supply chain since Embraer's privatisation. Import content has risen from approximately 68 per cent (Dagnino and Proença, 1989) in the 1980s to approximately 95 per cent (Bernardes, 2000a) in the 1990s, giving rise to questions regarding the participation of local Brazilian suppliers in the supply chain. Another question concerns the renewal of the Brazilian government efforts to build up a Brazilian aircraft manufacturing system of innovation (BASI) once it is assumed that there are weak linkages between firms and the S&T infrastructure as seen at national level. The next section attempts to improve understanding of these two issues.

4 The main features and changes in the BASI (1969–2002)

The BASI structure and characteristics of relationships are examined in this section. The structure of the Brazilian civil aircraft production and knowledge systems before the privatisation of Embraer (1969–1994) is outlined in section 4.1. The BASI is analysed in section 4.2 for the post-privatisation period, and, finally, changes in the structure and collaborative agreements are examined in section 4.3.

4.1 The Brazilian civil aircraft production and knowledge systems in the pre-privatisation phase (1969–1994)

The structure of the Brazilian civil aircraft production and knowledge systems is shown in figure 4.2. The production system generally consists of civil aircraft design and production, and four supplying tiers. Embraer manufactured a range of small-bodied aircraft with 8–30 seats during the period 1969–1994. Approximately 68 per cent of tier one suppliers were foreign firms, and supplied parts, components, structural parts, and systems. Embraer manufactured approximately 90 per cent of the fuselage structure of an aircraft during this time. Information about the type of goods and services the local firms provided is lacking, although some interviewed firms were supplying seats, landing gear parts, and engine parts and components, in addition to other parts.

The actors in the knowledge system were mainly (Bernardes, 2000a; Dagnino and Proença, 1989; Frischtak, 1992, 1994):
(i) CTA – helping the aircraft producer and Brazilian suppliers to develop product and process technologies.
(ii) NASA – with a procurement agreement with Embraer to transfer information on wing sections, and with agreements with the CTA.
(iii) Embraer – developing design, product, and process capabilities. The company invested in the acquisition of engineering competence in many fields (mechanics, electronics, and materials, etc.). Engineering information was also transferred and Brazilian suppliers trained in their weaker fields.
(iv) An American firm – with a collaborative agreement with the aircraft producer, which included support to Embraer for developing marketing and production capabilities.
(v) An American and an Italian firm – with a collaborative agreement for the development of a defence programme with the aircraft producer that included transfer of manufacturing process technologies.
(vi) Other Brazilian and foreign universities – the aircraft producer had many agreements with Brazilian and foreign universities for training highly qualified engineers in research skills.

The collaborative agreements in the BASI were basically procurement, licence agreements for technology transfer, and technological collaboration. Procurement agreements between Embraer and Brazilian local suppliers included technical support for adopting new technologies mainly in defence programmes. The Ministry of Defence played an important role in coordinating and negotiating Embraer's procurement agreements, mainly because the company played a strategic role in the security of the country (Green, 1987). Therefore, all procurement agreements were strategically negotiated by the Ministry of Defence and included technology transfer from important foreign suppliers to Embraer and from Embraer to local suppliers. The technology transfer contracts (licence agreements) were under the Brazilian government offset policies.

The technological collaboration was characterised by informal contracts between the aircraft producer and the CTA, and formal contracts with foreign universities and competitors, for example, Piper/USA (Frischtak, 1992; Green, 1987). These were also coordinated by the Ministry of Defence.

At the beginning of the 1990s, the Brazilian Ministry of Defence discontinued all its projects in the aeronautical field. It was a period of great macro-economic recession and the Brazilian government cut expenses in many areas including procurement of training aircraft, and research and development related to defence. The aircraft producer reduced the number of employees to approximately 70 per cent and production activities to approximately 50 per cent (Bernardes, 2000a).

Embraer's financial crisis had an impact on its supply chain and major restructuring occurred from 1990 to 1994. Many Brazilian supply firms exited the market, while others were founded by ex-employees of Embraer. Changes in the production system and knowledge system observed after 1994 are explained in the next section. As a consequence of these changes, the capabilities of local suppliers which had not exited the market have been questioned as regards their ability to manage and generate the necessary changes in order to maintain and improve their market position.

4.2 The BASI after privatisation (1995–2002)

Important changes occurred in the 1995–2002 period consequent on the privatisation of Embraer. The new Brazilian government industrialisation policies (i.e. liberalisation policies) have also had an impact on the sector. The changes in the production and knowledge systems are explained below. It is important to notice that the main aircraft models changed from turboprops to modern regional jets with higher passenger capacity and better performance (table 4.1). According to interviews, the main regional jet programmes are the ERJ 145 and the ERJ 170.

The production system is composed approximately 95 per cent of foreign firms, which produce structural parts (front, central and forward fuselage, wings, etc.) and systems (avionics, power, etc.) in the first tier of the supply chain. The few other products manufactured in Brazil are made by Embraer itself and by some subsidiaries of foreign first tier firms. There are more local firms within the tier two suppliers, which deal with metal processing, engineering projects, and software engineering services (Bernardes, 2000a,b).

The structure of the knowledge system has increasingly spread outwards since the beginning of the 1990s, because customers as well as a few first tier suppliers have participated greatly in product development. However, the most important actors have been: (i) local S&T institutes and universities, such as the CTA, ITA, University of São Carlos, University of São Paulo, etc.; (ii) foreign S&T institutes mainly located in the United States, Europe, and Russia; (iii) Embraer; and (iv) foreign suppliers mainly located in the United States, Europe, and Japan. Technological knowledge development concerns the application of basic knowledge from foreign S&T institutes for local technological improvement. These improvements are frequently associated with the product, the technological upgrading of the process, operational procedures, and

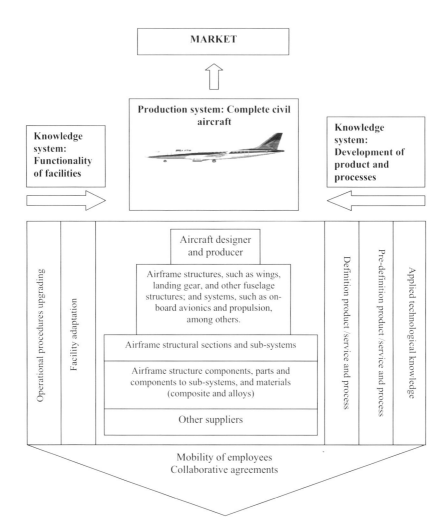

Figure 4.2 Knowledge and production systems in the Brazilian civil aircraft manufacturing system of innovation.
Source: Author.

facilities production. The two different types of technological knowledge are closely associated, as explained in section 2.

The difference between the general civil aircraft manufacturing system (fig. 4.1) and that in Brazil (fig. 4.2) basically relates to the knowledge system function. Specifically, the BASI has three broad important characteristics: (i) technological knowledge flows for the application of basic knowledge and technologies from developed countries (Bernardes, 2000a); (ii) the actors are mainly foreign, although CTA and Embraer are extremely important; and (iii) formal and informal collaborative agreements are supporting technological upgrading of Brazilian firms, which may result in incremental innovations in local firms which have reached the limit of their efforts as regards knowledge accumulation and learning processes. Local S&T institutes and universities in partnership with Embraer carry out applied technological research. Embraer has agreements also with foreign S&T institutes for applied research. There are also important technological knowledge flows in to the BASI from tier one foreign suppliers of structural parts and systems.

Tier two local suppliers have also contributed to local technological knowledge development according to the complexity of their products: local engineering project and software engineering firms are more integrated in the knowledge system than local metal processing firms. Therefore, Embraer and tier one foreign firms are the most integrated in the knowledge system and consequently transmit the new technologies down the supply chain.

There are at least five types of collaborative agreements in the BASI: (i) joint venture; (ii) risk-sharing; (iii) risk-sharing and co-design; (iv) preferred supplier arrangement; and (v) technology collaboration. An example of a joint venture is Eleb, a spin-off from a division of Embraer which was created by Embraer and a foreign supplier in 1999. The company manufactures parts and components for hydraulic systems and landing gear for medium body aircraft (35–108 seats).

Embraer and four foreign suppliers signed a "risk-sharing agreement" to develop the ERJ 145 jetliner. This agreement involves technological knowledge transfer for production in proportion to the partner technological capabilities and according to contract particulars. It includes, for example, engineering consultancy and assistance from Embraer to partner firms. The four partners took shares in the development costs of the ERJ 145 aircraft and in its revenues in proportion to their investments.

A "risk-sharing and co-design agreement" has been signed between Embraer and approximately 20 tier one foreign suppliers for sharing the development costs, revenues in proportion to each partner's investment, joint product and process definitions, and operational procedures of the

ERJ 170 project. This agreement involves formal and informal technological knowledge transfer between partner firms that have engineers working at Embraer's plant in Brazil. Technological transfer is limited to that necessary to perform the programme and restrictive contractual clauses have been agreed between the partners. Each partner has limited information about the overall ERJ 170 aircraft project, which is known only by Embraer.

"Preferred supplier arrangements" demand less commitment between the parties than the other three agreements explained above. They have been developed between Embraer, foreign suppliers, and tier two local suppliers, and between tier one foreign suppliers and tier two local suppliers. They are based on short-term procurement agreements that include engineering knowledge transfer depending on the engineering capabilities of the parties, and are embodied in technical specifications (blueprints) and technical assistance. These agreements can be extended to the long term if the supplier performs satisfactorily in reducing costs, improving quality, and developing technologically. There are cases where tier two local suppliers are working to develop engineering projects inside Embraer's plant and a tier one foreign supplier's plant. These cases presuppose a high level of commitment of partner firms to confidentiality and limited technological knowledge transfer. However, they are generally related to servicing "mature" technologies or those that are not considered strategic for Embraer or tier one foreign partners.

There are "technological collaborations" between the aircraft producer and local and foreign S&T institutes and universities in the BASI. These collaborations are concerned with: (i) the development of applied research in specific technological areas; (ii) servicing tests, experiments, etc.; and (iii) the training and qualification of engineers and technicians. This type of collaboration is joint funded by the Brazilian government when it involves local S&T institutes and universities and Embraer. The results of these collaborations benefit the Brazilian civil aircraft production system depending on the local firm's technological capabilities. The main academic and political concern is related to the lack of local firms' technological capabilities to absorb new technological knowledge and improve their technological capabilities so as to compete in international markets.

4.3 Changes in the BASI during the 1990s

The Brazilian civil aircraft production and knowledge systems have changed since the privatisation of the aircraft producer in 1994. These changes are examined by comparing the PS and KS in the two most important civil aircraft projects since 1994: (i) the ERJ 145 jetliner; and (ii) the ERJ 170 jetliner.

4.3.1 ERJ 145 jetliner: production and knowledge systems

The main characteristics of the PS for developing the ERJ 145 jetliner are:
- Embraer has approximately 450 tier one suppliers (Bernardes, 2000b).
- The tier one suppliers are divided into producers of structural parts, systems, and parts, components, and sub-systems.
- Local tier two firms are suppliers of metal processing, engineering projects, and software engineering services to the aircraft producer.
- There is a risk-sharing agreement between Embraer and a few tier one foreign suppliers of structural parts. They share the development costs of the programme and its revenues in proportion to their investments. Technological knowledge improvements depend on each party's own development. This partnership lasts for the life of the aircraft, which will be approximately 30 years: it is unlikely that re-sourcing will occur.
- Approximately 90 per cent of the tier one suppliers are foreign firms, with Embraer itself being the main tier one supplier in Brazil.

The main actors in the KS are Embraer and the CTA. Specifically, they research the application of basic knowledge created in developed countries. However, the national knowledge system has incomplete assets to support applied technological knowledge development in this industry. Further sources of technological knowledge are foreign S&T institutions, foreign tier one suppliers, and other organisations. The technological knowledge acquired is transmitted down the local supply chain through technological collaboration between Embraer, local S&T institutions, local suppliers, and the government; and through preferred supplier arrangements between Embraer and local suppliers. The local suppliers have adopted the new technologies to the extent of their technological capabilities.

4.3.2 ERJ 170 jetliner: production and knowledge systems

The main characteristics of the PS for developing the ERJ 170 jetliner are:
- Embraer has approximately 20 tier one suppliers in this programme (Bernardes, 2000b).
- Tier one suppliers are divided into producers of structural parts and systems.
- Local tier two firms supply metal processing, engineering projects, and software engineering services to Embraer and foreign tier one suppliers.
- Some tier one foreign suppliers have set up plants in Brazil.

- There is a risk-sharing and co-design agreement between Embraer and tier one foreign suppliers. They have shared both development costs and development activities. There are teams of engineers from each partner firm developing their corresponding structural parts or systems at Embraer's plant in Brazil, where they remain until aircraft certification. They are working together with Embraer's engineers to detail the product design, functionality, and performance of the ERJ 170. These partnerships will last for approximately 30 years as in the ERJ 145 project.

The main actors in the KS for developing this programme are Embraer, partner tier one foreign suppliers, and foreign S&T institutes. The participation of the CTA involves servicing tests, experiments, and the training of engineers. There is an extended structure for developing applied research in many areas, which include other national and foreign S&T institutions. Technological knowledge is diffused to local tier two suppliers not only by the CTA and Embraer, but also by tier one foreign suppliers.

These changes in the production and knowledge systems have raised questions about the consolidation of the Brazilian tier two suppliers' technological capabilities. Specifically, the questions are concerned with the development of knowledge-changing activities within the local supply chain and underlying learning processes. Thus, the next section presents some concluding remarks about the BASI and attempts to highlight some important questions that should be addressed in further research.

5 Concluding remarks

The findings outlined above point to two sets of changes in the BASI. The first set is related to structural changes in the production system, which are:
- The increased participation of Embraer in the world civil aircraft market.
- The localisation of subsidiaries of tier one foreign suppliers in Brazil.
- The expansion in the possibilities of local tier two firms supplying tier one foreign suppliers.

The second set of changes is related to the BASI knowledge system, namely:
- A decrease in the role of the CTA as one of the main actors in the national knowledge system.
- An increase in the role of foreign suppliers and foreign S&T institutes in technological knowledge transfer to the Brazilian civil aircraft production system.

- An increase in the role of the Brazilian knowledge system in supporting the application of basic technological knowledge developed abroad to the local production system demands.

The KS has developed a more extended configuration in the post-1994 period. Embraer and the CTA are no longer nearly the only sources of technological knowledge for the Brazilian local suppliers now that localising foreign tier one suppliers are relating closely to local firms as well as other S&T institutes (national and foreign).

What therefore can we conclude from this overall picture about the function of the BASI post-1994? At first glance, there are two possible answers that should be examined in further research. First, the local Brazilian tier two suppliers and the aircraft producer are increasing both their technological capabilities and their participation in the highly competitive world civil aircraft market. However, the participation of the CTA as a *supplier* of applied technological knowledge is decreasing. In addition, the Brazilian civil aircraft production and knowledge systems have spread outwards looking for the best suppliers of products and applied technological knowledge for supporting their requirements to compete in the world market.

Second, tier two local suppliers are not increasing their technological capabilities but have just become better at using their technological knowledge base in order to maintain or adapt themselves to the actual market requirements. The role of the CTA is much more important for these firms than it seems because it is supporting firms by training engineers and technicians, and servicing tests, experiments, and so on. It is probably the case that even Embraer is maintaining and adapting the technological knowledge base developed before 1994 through its close integration in the international civil aircraft production and knowledge systems. Also, the Brazilian civil aircraft production and knowledge systems have changed the chief actors but maintain their principal features as adopters of aeronautical engineering knowledge.

In any case, the question raised is whether the first or second hypothesis is more valid: Have tier two local firms developed their technological knowledge base through knowledge-changing activities, by adapting it through knowledge-adapting activities, or by maintaining it through knowledge-using activities? The answer to this question has many policy implications, at least two of which are very important for the consolidation of Brazilian civil aircraft manufacturing. First, the Brazilian government should strengthen the role of the local S&T institutes in supporting the diffusion and utilisation of applied technological knowledge within the BASI. The policy instruments should include increasing investment in training of researchers, highly qualified engineers, and technicians and the development of applied aeronautics research. Second, interaction should be encouraged between tier two local suppliers, national S&T institutes, and the international civil aircraft production and knowledge systems. The results of these policies should be increasing economies of scale, applied technological knowledge acquisition, and upgrading of technological capabilities. These three possible results from policies for production and knowledge development are very important for the consolidation of Brazilian civil aircraft manufacturing worldwide.

Concluding, the case study of the BASI should be used by academics and policy makers worldwide as an example of successful government policy for developing design, production, and marketing capabilities in the manufacturing of a high tech product such as an aircraft. However, the effect of a strong final manufacturer on local sectoral development is still unclear. It may stimulate technological upgrading in the product, the production process, operational procedures, and facilities in local suppliers, and development of applied research in areas such as new materials and information technologies, etc., that may spill over to other sectors. However, local suppliers, which in the case of Brazil are small firms, are heavily dependent on government incentives to continue and improve their position in a very competitive and international market; local small firms compete not only among themselves but also with foreign firms. Aircraft manufacturing is characterised by few economies of scale, high costs, and long-term return on investments, and is heavily affected by international economic crises. Therefore, small firms experience many barriers to developing themselves and need government support in areas such as finance, engineering education, and development of specific networks and mechanisms to stimulate technological knowledge transfer considered fundamental to small firms in order to explore the benefits of international production and knowledge systems.

Notes

I would like to thank Professor Sunil Mani (UNU-INTECH), Roberto Fontana (Bocconi University), Marcia Darós (UNU-INTECH), Renato Dagnino (UNICAMP/DPCT), and Aaron Griffiths for their detailed comments on an earlier draft. I am also grateful for valuable discussions with my supervisor, Mr. Martin Bell, and my colleague André Campos (SPRU).

This paper is the result of a paper presented at the Fourth UNU/INTECH-CERES WP3/EADI Conference on Innovation, Learning and Technological Dynamism of Developing Countries. However, the views expressed herein are those of the author.

1. The CTA is composed of four institutes: the Technological Institute of Aeronautics (ITA), the Institute of Aeronautics and Space, the Institute of Advanced Studies, and the Institute for Development and Coordination of the Aerospace Industry (IFI). The most important Brazilian institution training aeronautical engineers is ITA (http://www.cta.br; accessed on 18 February 2002).
2. The family or commonality concept in aircraft design means that a given aircraft, such as

the Boeing 727, generates a succession of modified designs, frequently through stretching the fuselage, with similar operation characteristics (Mowery and Rosenberg, 1985).
3. The first and second largest civil aircraft producers are, respectively, Boeing from the US and Airbus from Europe.
4. Aerospatiale Matra, 5.67 per cent; Dassault Aviation, 5.67 per cent; Thomson-CSF, 5.67 per cent; and Snecma, 2.99 per cent (report in Bernardes, 2000b).
5. Information gathered from website: http://www.embraer.com.br/.

BIBLIOGRAPHY

AIAB (Associação das Indústrias Aeroespaciais do Brasil) (2001) "O Setor Aeroespacial Brasileiro: Corte em 2000" (The Brazilian Aerospace Sector in 2000), mimeo, São José dos Campos (São Paulo): Brazilian Aerospace Industrial Association.

Arnt, R. (2001) "Fim da Cópia? O Tempo dos Pacotes Fechados já Passou", *Revista Exame* 22 August: 54–58, São Paulo: Editora Abril.

Bell, M. and M. Albu (1999) "Knowledge Systems and Technological Dynamism in Industrial Clusters in Developing Countries", *World Development* 27(9): 1715–1734.

Bell, M. and K. Pavitt (1993) "Technological Accumulation and Industrial Growth: Contrasts between Developed and Developing Countries", *Industrial and Corporate Change* 2(2): 157–209.

Bernardes, R. (2000a) *Embraer: Elos entre Estado e Mercado*, São Paulo: Editora Hucitec/FAPESP.

—— (2000b) *Oportunidades de Mercado, Produção e Acesso ao Conhecimento: Linhas de Ação Para o Fortalecimento da Performance Tecnológica do Setor Aeronáutico*, Rio de Janeiro: FINEP.

—— (2001) "Articulação e Especialização Produtiva de MPMES: O Caso do *Cluster* Aeronáutico da Região de São José dos Campos, São Paulo" (Articulation and Productive Specialization of SMEs: Case Study of the Aeronautic Cluster in the Region of São José dos Campos, São Paulo), first draft report to ECLAC/UN, mimeo, São Paulo: Fundação SEADE.

Carlsson, B. (1995) *Technological Systems and Economic Performance: The Case of Factory Automation*, Dordrecht: Kluwer.

Carlsson, B. and G. Eliasson (1993) "The Nature and Importance of Economic Competence", selected papers relating to the presentations given at the "Swedish Seminar" held at University of Sussex, Brighton: SPRU, University of Sussex.

Carlsson, B. and S. Jacobsson (1993) "Technological Systems and Economic Policy: The Diffusion of Factory Automation in Sweden", selected papers relating to the presentations given at the "Swedish Seminar" held at University of Sussex, Brighton: SPRU, University of Sussex.

—— (1994) "Technological Systems and Economic Policy: The Diffusion of Factory Automation in Sweden", *Research Policy* 23(3): 235–248.

Cassiolato, J. E. and H. M. M. Lastres (1999) "Local, National and Regional Systems of Innovation in the Mercosur", paper presented at the DRUID

Summer Conference on National Innovation Systems, Industrial Dynamics and Innovation Policy, June 9–12, 1999, Rebild.

Cimoli, M. (2000) "Economic and Technological Shocks: The Dynamic of Innovation Systems and Networks – The Case of Latin American Countries", paper presented at the Conference on The Other Canon in Economics, August 2000, Oslo.

Cooke, P., M. Uranga and G. Etxebarria (1997) "Regional Innovation Systems: Institutional and Organisational Dimensions", *Research Policy* 26(4/5): 475–491.

Costa, I. and S. R. R. Queiroz (2002) "Foreign Direct Investment and Technological Capabilities in Brazilian Industry", *Research Policy* 31(8–9): 1431–1443.

Coutinho, L. and J. C. Ferraz (1993) "Estudo da Competitividade da Indústria Brasileira: Competitividade da Indústria Aeronáutica", Financiadora de Estudos e Projetos (FINEP), Programa de Apoio ao Desenvolvimento Científico e Tecnológico (PADCT), Campinas: Ministério da Ciência e Tecnologia (MCT).

Dagnino, R. (1993) "Competitividade da Indústria Aeronáutica", Nota Técnica do Complexo Metal-Mecânico, Estudo da Competitividade da Indústria Brasileira, online at: http://www.mct.gov.br (accessed in 2001).

Dagnino, R. and D. J. Proença (1989) *The Brazilian Aeronautics Industry*, Geneva: International Labour Office.

Dahlman, C. and Frischtak, C. (1990) "National Systems Supporting Technical Advance in Industry. The Brazilian Experience", mimeo, Brighton: SRPU, University of Sussex.

De Ferranti, D. and G. E. Perry (2002) *Closing the Gap in Education and Technology*, World Bank Latin American and Caribbean Studies, Washington, D.C.: The World Bank.

Donângelo, A., F. P. Coelho and J. Ichimura (2000) "Gerência do Marketing Mix Internacional de uma Empresa Brasileira: O Caso da Embraer" (Management of International Marketing Mix in a Brazilian Company: The Case Study of Embraer), *Cadernos Discentes COPPEAD* 2: 39–55, Rio de Janeiro: COPPEAD/UFRJ.

Edquist, C., ed. (1997) *Systems of Innovation: Technologies, Institutions and Organizations*, London: Pinter.

Ellis, B. W. C., T. Williams, M. J. Gregory and R. Maull (2000) "Driver and Response Framework within the Global Aerospace Manufacturing Industry", in Cambridge Centre for International Manufacturing, ed., *International and Strategic Network Development: The Proceedings of the 5th International Manufacturing Research Symposium, 3rd–5th September 2000*, Cambridge, UK: Churchill College, Cambridge University, pp. 85–104.

ESPM (2001) "A Nova Face da Gerência de Projetos – O Projeto do Avião ERJ-170 na Embraer" (The New Face of Project Management – Project of the Aircraft ERJ-170 at Embraer), *Revista ESPM* (March/April), Special Edition: 50 years, Brazil: São Paulo, pp. 89–95.

Fong, P. (2001) "The Dogfight for Bombardier's Niche: The Canadian Jet Maker Is Battling Embraer and Boeing?", *Business Week* (European Edition) 3728–1058 (6 August): 6.

Freeman, C. (1987) *Technology and Economic Performance: Lessons from Japan*, London: Pinter.
Frischtak, C. (1992) *Learning, Technical Progress and Competitiveness in the Commuter Aircraft Industry: An Analysis of Embraer*, The World Bank Industry and Energy Department, OSP: 68, Washington, D.C.: The World Bank.
—— (1994) "Learning and Technical Progress in the Commuter Aircraft Industry: An Analysis of Embraer's Experience", *Research Policy* 23(5): 601–612.
Gelsing, L. (1992) "Innovation and the Development of Industrial Networks", in B. A. Lundvall, ed., *National Systems of Innovation: Towards a Theory of Innovation and Interactive Learning*, London: Pinter, pp. 116–128.
Goldstein, A. (2001) *The Political Economy of High-Tech Industries in Developing Countries: Aerospace in Brazil, Indonesia and South Africa*, Paris: OECD Development Centre, online at: http://www.oecd.org.
Green, R. D. (1987) *Brazilian Government Support for the Aerospace Industry*, Washington, D.C.: US Department of Commerce/International Trade Administration.
Hwang, C. (2000) "The Aircraft Industry in a Latecomer Economy: The Case of South Korea", academic thesis, Brighton: SPRU, University of Sussex.
Katz, J. (2000) *Passado y Presente del Comportamiento Tecnologico de America Latina* (Past and Present of the Technological Behaviour in Latin America), Revista Serie Desarrollo Productivo, number 75, Santiago de Chile (Chile): Red de Reestructuracion y Competitividad, Division de Desarrollo Productivo y Empresarial, ECLAC, United Nations.
—— (2001) "Structural Reforms and Technological Behaviour: The Sources and Nature of Technological Change in Latin America in the 1990s", *Research Policy* 30(1): 1–19.
Kim, L. and C. Dahlman (1992) "Technology Policy for Industrialisation: An Integrative Framework and Korea's Experience", *Research Policy* 21(5): 437–452.
Lundvall, B. A. (1992) *National Systems of Innovation: Towards a Theory of Innovation and Interactive Learning*, London: Pinter.
Malerba, F. (2002) "Sectoral Systems of Innovation and Production", *Research Policy* 31(2): 247–264.
Mani, S. (2001) "Government, Innovation and Technology Policy, an Analysis of the Brazilian Experience During the 1990s", Discussion Paper Series, 2001-11, Maastricht: United Nations University, Institute for New Technologies, online at: http://www.intech.unu.edu (accessed on 22 February 2002).
Modesti, A. (2001) "Compensação Comercial, Industrial e Tecnológica: Offset no Brasil" (Commercial, Industrial and Technological Compensation: Offset in Brazil), mimeo, São José dos Campos: Centro Técnico Aeroespacial (CTA), IFI, Nucleus of the Industrial Development Division.
Mowery, D. and N. Rosenberg (1981) "Government Policy and Innovation in the Commercial Aircraft Industry, 1925–75", mimeo, Brighton: SPRU, University of Sussex.
—— (1985) *The Japanese Commercial Aircraft Industry since 1945: Government Policy, Technical Development, and Industrial Structure*, Stanford: Stanford University, The International Strategic Institute.

Nelson, R. and N. Rosenberg (1993) "Technical Innovation and National Systems", in R. Nelson, ed., *National Innovation Systems: A Comparative Study*, Oxford: Oxford University Press, pp. 3–21.

Padgett, T (2003) "Dogfight: Amid a Down Market in Regional Jets Brazil's Embraer and Canada's Bombardier Show Their Teeth", *Time Europe Magazine* 161(May 26): 21, online at: http://www.time.com/time/europe/magazine/article/0,13005,901030526-452785-1,00.html.

Quadros, R., A. Furtado, R. Bernardes and E. Franco (2001) "Technological Innovation in Brazilian Industry: An Assessment Based on the São Paulo Innovation Survey", *Technological Forecasting and Social Change* 67: 203–219.

Ruffles, P. (1992) "Partnerships and Sharing Risks – Making Other Peoples' Technology Work for You", mimeo, Brighton: SPRU, University of Sussex.

Steinmueller, W. E. (2000) "Knowledge and Learning in the Information Age", paper prepared for the DRUID Summer Conference on The Learning Economy – Firms, Regions and Nation Specific Institutions, June 15–17, 2000, Rebild, online at: http://www.business.auc.dk/druid.

Todd, D. (1989) "The Industrialisation of the Aircraft Industry: Substance and Myth", Working Paper 29, Geneva: International Labour Office.

Vincenti, W. G. (1993) *What Engineers Know and How They Know It: Analytical Studies from Aeronautical History*, London: The Johns Hopkins University Press.

Viotti, E. B. (2002) "National Learning Systems: A New Approach on Technological Change in Late Industrializing Economies and Evidences from the Cases of Brazil and South Korea", *Technological Forecasting and Social Change* 69(7): 653–680.

Wörner, S. and T. Reiss (1999) "Technological Systems", background briefing paper 2, in J. Senker, O. Marsili, S. Wörner, T. Reiss, V. Mangematin, C. Enzing and S. Kern, eds., *Literature Review for European Biotechnology Innovation Systems (EBIS)*, Brighton: SPRU, University of Sussex, pp. 36–52, online at: http://www.sussex.ac.uk/spru (accessed on 4 December 2001).

Yu, A. S. O. B., J. Francisco, P. T. Nascimento, A. S. Camargo Jr. and M. Takami (2001) "Desenvolvimento de Produto e Processos: Um Estudo de Caso do ERJ 170", III Congresso Brasileiro de Gestão de Desenvolvimento de Produto, conference CD, Florianópolis (Brazil): Universidade de Santa Catarina.

5

The political economy of technology policy: The automotive sector in Brazil (1950–2000)

Effie Kesidou

1 Introduction

This paper starts from the widely held view that technological development of the industrial sector in developing countries can be promoted by purposive state intervention, and that technology policy[1] is an important instrument to achieve this. Currently, there are two dominant branches of research on technology policy. The first approach, popular among economists, mainly concerns itself with the reasons why technology policy is justified. Central importance is given to market failures (Amsden, 1989; Wade, 1990; White, 1988).[2] The second approach, which is common among political scientists, puts primary emphasis on (political) conditions that make state intervention effective (Chang, 1994; Evans, 1995).

The main argument developed in this paper is that both these approaches have certain shortcomings. They do not have much to offer when one wants to understand (i) how technology policy is actually generated and shaped in particular countries, and (ii) the reasons why certain policies succeed or fail in a particular context. The main problem is that the two above-mentioned approaches yield only partial insights. The market-failure approach adopted by economists is concerned mostly with justifying technology policy, and less with identifying the factors that influence its actual formation and content, and how that bears on success. Conversely, the approach used by political scientists is focused on a state's capabilities in undertaking technology policy and how this impacts on effectiveness, but ignores the economic factors that shape the policy.

The formation and evolution of technology policy is a complex process influenced by many factors. By unravelling these factors, more light is shed on its prospective impact on the technological dynamism of industry. This paper aims to make a contribution on this subject. To this end, a new framework for analysing technology policy is developed, which places technology policy in its specific historical and institutional context, and conceptualises it as resulting from interaction between agents and the international context. The new framework incorporates the power relations associated with both political and economic factors, and examines their impact on the formation of technology policy.

The framework is applied to a case study of the Brazilian automotive industry. The industry is of particular interest as it has been an engine of domestic expansion for a prolonged period of time. During the 1960s in particular, it supported increases in employment, investment, and incorporation of skills and technology in the context of import substitution industrialisation efforts. Industrial and technology policies were the main tools used by the Brazilian government to transform the country from an agricultural to an industrial economy. Technology policy became a key objective in the 1970s when the National System of Scientific and Technological Development (SNDCT) was created. The plan sought to develop the country's physical and human R&D infrastructure, to subsidise financing of local firms in order to stimulate the development of technological capabilities, and to purchase foreign technology.

The focus of the case study is on the technological dynamism of the auto industry, defined in a broad sense. In the context of a developing country like Brazil, technological dynamism cannot be assessed merely by considering R&D investments (as is commonly done in studies focused on developed countries), since the dominant pattern of technological development and industrial growth consists of assimilation of foreign technologies. Hence, technological dynamism in this context should also include such elements as technological strengthening and modernisation of production operations, expansion of production capacity, upgrading of quality, human capital formation, development of export capacity, and creation of domestic linkages.

This chapter is organised as follows. The second section presents a review of the two main branches of research in the literature on technology policy and elaborates their important weaknesses. The third section introduces the new framework, which addresses these problems. Section 4 explores the technology policy that has been implemented in the Brazilian automotive industry and evaluates its effects on technological dynamism, using the new framework. Section 5 presents the main conclusions of the research.

2 Theories of technology policy: A review

The purpose and the role of technology policy can be understood by analysing the two main approaches that have been developed to study the phenomenon. In particular, this will focus on the validity of the conception of the state as the sole formulator of technology policy. Although most scholars associated with the two approaches share the broad view that governments play a certain positive role which is more or less in line with the "developmental state" approach,[3] different analytical methods are used, and the approaches tend to serve different purposes.

2.1 Contributions made by economists

Economists justify state intervention on the grounds that market failures are more pervasive than a laissez-faire stance assumes. Markets are characterised by imperfections, particularly in LDCs (less developed countries). Therefore, it is argued that governments should have an active role in trying to eliminate market failures and foster industrial development through technology policy.

2.1.1 The protectionist and the coordinating states: The infant industry argument

One of the most notable reasons advanced in favour of protection is known as the infant industry argument. The central idea of the infant industry principle is that LDCs, in order to induce industrialisation, need to give temporary assistance,[4] in the form of protection from higher productivity producers, to the newly emerging manufacturing sector.

The inefficiency of infant industries stems from a lack of production experience caused by late investment. The reasoning behind this argument can be analysed by considering "dynamic increasing returns" or "dynamic economies".[5] The dynamic economies arise because technology is not perfectly tradable. According to Lall: "Technological knowledge is not shared equally among firms, nor is it easily imitated by or transferred across firms. Transfer necessarily requires learning because technologies are tacit" (Lall, 1992: 166). Since technological knowledge is not easily acquired and "first movers" have a significant cost advantage, it makes sense to protect latecomers. They should be given time to "grow up" and "learn-by-doing".

The argument for protection of infant industries can be further supported in the case of positive inter-firm effects or externalities. First, the diffusion of knowledge is the most important technological externality that affects industrial development. That is, due to labour mobility or imitation, the benefits of innovation are not entirely captured by the innovator. Second, pecuniary externalities are relevant to industrial development when they are reciprocal and the industry is subjected to increasing returns. This can lead to "coordination failure" (Corden, 1974). This implies that prices are a poor guide for future investment since they do not reflect potential reciprocal externalities. This kind of imperfection calls for government involvement in diffusing information and in coordinating investment decisions by the private sector.

Thus, two reasons have been put forward for protecting an infant industry: first, to stimulate the learning effects that will improve productive efficiency; and second, these learning effects might spill over into the rest of the economy as managers and workers open new businesses or move to other industries in the economy.

2.1.2 The entrepreneurial state

The "entrepreneurial state" argument calls for a more extensive role for the state in the private sector than the protectionist state argument. In order to enhance its international competitiveness, the state as an entrepreneur takes managerial decisions such as establishing new industries and setting incentive structures, as well as taking responsibility for reshaping the domestic industrial structure with strategically targeted industries.

For example, it has been argued that East Asian newly industrialised countries (NICs) have assisted the private sector in building competitive assets such as professional management capability and large-scale corporations (Wade, 1990). In order to socially construct these competitive assets for productive purposes, governments deliberately distorted prices. Justification lies in a different methodological approach which views market failures in production (as opposed to exchange) as central to industrial development (Amsden, 1997). Therefore, economic development does not require the reduction of market failures, but economic development is a process of "building them" in the production sphere (Amsden, 2001).

The above argument underlines a fundamental difference between "the protectionist" and "the entrepreneurial" views of the state. While the protective state argument justifies temporary protection in order to correct market failures, the entrepreneurial state argument supports intervention in resource allocation for strengthening market failures and creating "competitive assets".

The contributions made by economists to the role of the state in technology policy are valuable, as they underline the justification for a state's involvement in industrial development.

2.2 Contributions by political scientists

Researchers writing from the point of view of political science emphasise political conditions that make state intervention effective. The nature of the state structure and its social ties to dominant classes are of significant importance.

2.2.1 Embedded autonomy

The most noteworthy theme in the political approach towards technology policy is state autonomy. According to Evans (1995), a well-developed and cohesive bureaucratic apparatus based on a long-term process of institution building is a prerequisite for effective state action. The second requirement is a state being autonomous from the dominant class. Finally, Evans introduces the term "embeddedness" to express the characteristics that a state should possess. According to his view, the state should selectively stimulate, complement, and reinforce entrepreneurship. In order for a state to be effective in this role, it should not be insulated from the society. Rather, the state should be embedded in a set of social ties to impose policy goals effectively on private economic growth.

However, this theory does not explain whence state autonomy is derived, or which political-economic factors determine state autonomy. A contribution to fill this gap has been made by Chang (1994), who refers to South Korea's example of the military regime, arguing that the political nature of the regime determines whether a state is strong and therefore autonomous. However, Chang's argument is not universally applicable. For example, Brazil also had a military regime for some time, but this regime did not exhibit the coherence and autonomy of South Korea. Thus, it appears that it is not always the nature of the political regime that determines the effectiveness of a state's policies.

To sum up, both the economic and the political science approaches have their limitations. Economists are preoccupied by arguments with which they can justify government intervention. However, they do provide broad pointers towards the conditions under which technology policy is likely to be beneficial, and the conditions under which it is likely to be counterproductive, but no more than that. This is because economists ignore the factors that influence the actual formation of technology policy, and the factors that impinge on the effectiveness of its implementation in practice. Political scientists, on the other hand, do address this problem to some extent, by considering the state's capabilities. However, they do not take account of the economic factors that shape technology policy. Thus, both approaches can at best provide partial insights.

3 A new framework for analysing technology policy

A comprehensive approach to the analysis of technology policy should be able to answer the following three questions: What are the forces that influence (the emergence of) technology policy? How is technology policy formed and (re)shaped? What is the impact of technology policy? Answering these questions sheds light on the reasons for the success or failure of technology policy.

We have seen that the extant approaches have largely neglected the second question concerning the forming and shaping of technology policy. Therefore, these approaches cannot claim to give a complete answer to the third question about the impact of technology policy. The new framework developed in the following paragraphs is designed to contribute to a better insight into these particular aspects which have remained underresearched within the existing approaches. The framework is meant as an instrument with which we can shed light on the factors that determine the emergence and shaping of technology policy and its evolution over time in specific settings. The framework addresses the specific content of the policies adopted, and issues bearing on the ways in which these policies are implemented.

In order to achieve this, the framework adopts a political economy approach. This approach is different from the monodisciplinary approaches discussed above, in that it tries to understand economic dynamics and the state in a societal context. Political economy approaches "try to uncover the patterns and trends of change in the relations between the state and society within a broad perspective comprising interest groups, social conditions and cultural factors" (Martinussen, 1999: 237). Menzies (1997) distinguishes three key dimensions of political economy analysis: (i) the "horizontal" dimension, which looks at competition as a process involving power relationships between actors having differing degrees of information; (ii) the "vertical" dimension, which refers to institutional structures governing the policy-making process in countries; and (iii) the "depth" dimension, which recognises the importance of history and time.

These three dimensions are reorganised somewhat for the purpose of the analysis in this paper, yielding three essential ingredients that make up the new framework. These are as follows. First, *the historical, political, and institutional characteristics* of the particular country are considered explicitly. Second, the *particular economic relations* (including power relations) *between different agents arising out of the particular industrial structure* are taken into account. Finally, the *impact of the international economic environment* on technology policy is considered. I discuss each in turn.

3.1 National historical and institutional characteristics

The success or failure of technology policy is strongly related to the particular historical, political, and institutional characteristics of the country where the policy is implemented (Amsden, 2001; Wade, 1990). For example, an important factor contributing to the success of East Asian economies has been the presence of human and physical capital.[6] The Japanese occupation of Korea and Taiwan left these countries with a developed infrastructure and a well-educated population. These factors played a fundamental role in industrial development, particularly as regards the adoption of new technologies. It is necessary to take the above characteristics into account in order to understand particular socio-economic relations and structures that affect technology policy. Neglecting the importance of initial economic conditions leads to erroneous perceptions of similarity, and thus also to the application of "golden policy prescriptions" across a wide variety of countries.

3.2 Political-economic relations between agents

Different forms of relations between economic agents and their specific political-economic interests influence technology policy directly or indirectly. These relations arise out of the industrial structure (Chang, 1994). Depending on the specific interests of economic agents and their relative bargaining power, technology policy is influenced and transformed. Having capital accumulation as a general guide to their actions, economic agents develop competitive or cooperative relations between themselves. The particular economic interests arising out of the above relations are of fundamental importance to the formation of technology policy.

3.3 The international environment

In the context of developing countries – even NICs – it is an oversimplification to consider governments or markets as entities that function separately from the international environment (as Evans did in his theory, for example). In an increasingly globalised world, it is naive to believe in the autonomy of government interests and power, or in the insulation of local markets from global market forces. Both governments and markets in developing countries are intricately linked to the external world. In particular, there are many forms of external dependence. On the other hand, we also cannot adopt the dependency theory (centre–periphery theory) that argues that there is a complete manipulation of developing countries by advanced industrial countries.

Here I put forward a more nuanced view, following for example Jenkins (1987), Dicken (1998), UNCTAD (2000), and Oman (1994). Developing countries can still exercise some discretion in policy-making. At the same time, developed countries lead the global economy in technological and innovative activities. In addition, the main research (academic or firm/industrial) institutes and the most important international organisations (World Bank, IMF) are based in developed countries. Their influence on drawing up and implementing policy in developing countries is crucial. Moreover, transnational corporations (TNCs), being strong players in developing country markets and having links in many places across the globe, can also influence technology policy in host countries directly and indirectly.

With the growth of TNCs, much international production takes place between affiliates of the same company (intra-firm trade). By the end of the 1980s, intra-firm trade constituted one third of international trade (Hewitt, Johnson and Wield, 1992). Therefore, many decisions about which countries produce which products on the international market are not brought about by impersonal market forces. Rather, they are administrative decisions, taken in the headquarters of major corporations. Thus, global constraints place compelling limits on what sectoral roles are possible. From minerals to autos, local sectoral strategies must continually contend with limits imposed by the way in which production and markets are structured globally.

Table 5.1 summarises the highlights of the extant approaches to technology policy, and contrasts these with the new approach outlined here. The new approach is complementary to the two existing approaches in that its main aim is to understand the factor that has been most neglected in the extant approaches so far, namely how technology policy emerges and takes shape, and how it evolves over time. In this way, the new

Table 5.1 Contrasting approaches to technology policy (TP): A summary

Approach focus	Economic	Political science	New framework
(1) Justification of TP	Main objective	Assumption	Assumption
(2) Generation/formation/ modification of TP	–	–	Main objective
(3) Effectiveness of TP	Secondary objective	Main objective	Secondary objective

framework should also be able to contribute to a deeper insight into the question of the effectiveness of technology policy.

4 Case study: The Brazilian automotive industry

The case study in this section explores technological policies and their effects on the technological development of the Brazilian automotive industry. It is divided into two parts. The first part looks at technology policy in car manufacturing from 1952 until 1974, while the second part discusses a deep shift in technology policy from 1975 until today. The main reason for such a division is that from the foundation of the industry (1952) until the economic "miracle" (1974) the state was very much involved in technology policy. However, it will be shown that the state was not the only factor in forming policy for the automotive industry. Other agents, such as parts manufacturers and TNCs, also contributed directly or indirectly to the decision-making. The second period, after 1975, is characterised by a gradual withdrawal of the state from technology policy. Nevertheless, decision-making has not been completely left to the invisible power of the market. States (in this case not the federal state) still influence technology policy, but TNCs are now the dominant party in terms of decision-making.

4.1 Strong state intervention in the Brazilian automotive industry (1952–1974)

4.1.1 Generation of the industry (1952–1961)

Initially, I address the fundamental question of why Brazil decided to produce automobiles in the first place. Then I use the framework outlined above to examine and explain the nature of the technology policies that were designed to this end, and identify the factors that influenced the nature of the adopted policies. Finally, the impact of these policies is assessed.

The establishment of car manufacturing was part of Kubitschek's "Target Plan", a state-sponsored industrialisation programme developed along the lines of import substitution industrialisation. The idea of a domestic auto programme first emerged in response to Brazil's post-war balance of payments crises,[7] but soon took on more strategic significance. Local production of motor vehicles, a major item in the import bill, was expected to ease foreign exchange constraints and attract foreign capital and technology into the country. This implies that initially import substitution of cars was not a technology policy aiming to develop an industrial base, but a fiscal policy intending to save foreign exchange.

The strong influence of the United States' model of industrial development led the auto industry to be viewed as a necessary element of an integrated industrial structure. Motor vehicles, with their extensive backward linkages, were expected to trigger the development of complementary industries. The auto plan was promoted strongly by Kubitschek, who in 1956 created the Executive Group for the Automotive Industry (Grupo Executivo da Indústria Automobilística, GEIA). This was the first step in the Brazilian government's shift towards an ambitious plan to develop a domestic automotive industry. They aimed to increase the local content in order to promote industrial transition from the simple assembly of imported completely knocked down kits to full-scale local manufacture of cars.

The basic element of the new policy was, first, to attract foreign capital and, second, to encourage companies to use domestic components and parts. This decision was driven by the intention of the government to create linkages within the domestic economy. The auto industry's demand for intermediate inputs such as steel, parts, and components would create pressure to develop other sectors in the economy. In addition, the lack of strong capital markets[8] meant that investment should be achieved by foreign direct investment.

In order to achieve the first goal, GEIA used indirect incentives to attract foreign investment for motor vehicle production. In general, the Kubitschek government relied on differential exchange rates for exports and imports to transfer resources to the industrial sector. In addition, fiscal incentives[9] and foreign exchange advantages[10] were provided by the National Bank for Economic Development (Banco Nacional de Desenvolvimento Económico, BNDE) to the 11 foreign companies[11] which participated in the plan. However, the above benefits were given only to firms that met the 90–95 per cent domestic content schedule.

Furthermore, the Brazilian government forced firms to increase the level of local content by closing the market to imports. Tariff protection of goods produced domestically became generalised in what is called the Law of Similars.[12] Many foreign investors, under the threat of this law and the fear of being excluded from the Brazilian market, moved from importing cars into assembly of parts/components or from assembly into manufacturing of automobiles using mainly local parts/components (Gordon and Grommers, 1962).

The effects of the "Law of Similars" were twofold: first, it imposed pressure upon foreign firms to make use of domestic parts and components; and, second, it supported the assembly/parts industry by providing access to a large protected market.

A combination of economic and political factors contributed to the successful establishment of automotive manufacturing in Brazil, which we can study by adopting the framework developed in the previous section. Following the three key elements of this framework, the most relevant factors in the Brazilian case are: (i) the institutional and administrative capacity of the Brazilian state; (ii) the politico-economic relations between the state and the parts manufacturers arising out of the industrial structure; and (iii) the international environment. Each factor will be discussed in turn.

Regarding the characteristics of the Brazilian state, it has been observed that Brazil exhibits a degree of bureaucratic organisation, although not the degree of corporate coherence enjoyed by strong states such as South Korea. The Kubitschek government was an elected, civilian government, but depended on the cooperation of the landed oligarchy. However, the success of the automotive implementation project was based on the creation of efficient institutions such as GEIA and BNDE to coordinate and finance the project. Evans (1995) has called those institutions that provided the basis for successful projects of sectoral transformation "pockets of efficiency". For example, GEIA had sufficient authority and coherence to pose a credible threat of market closure to foreign firms. Moreover, it compensated for the increased cost and risk associated with investment and the high degree of local requirement by subsidising foreign firms.

Accordingly, the successful implementation of car manufacturing in Brazil was due to the capacity of the government to build efficient institutions. Those institutions were the main players, in that they formed policy and influenced private sector behaviour. Therefore, the government's role was crucial to the entry of foreign firms into the Brazilian market. If we take into account Shapiro's (1994) hypothesis that "no firm planned to manufacture vehicles in Brazil in the absence of government intervention", the role of the state in attracting foreign capital and establishing the industry was fundamental. Next, I turn to the politico-economic relations between the state and the parts manufacturers. Local parts producers played an essential role in forming Brazil's auto-assembly industry. It was not only the strong state that took the decision to implement a domestic automotive industry. The decision was shaped under the pressure of a strong lobby of auto components manufacturers. Local firms, which had developed to supply the replacement market and produce some parts for the assemblers, viewed the massive entry of foreign capital not as a threat but as the creation of new investment opportunities. Therefore, these companies influenced government to draw up legislation such as the "Law of Similars" that deliberately created "hybrid practices" instead of mass production, including relationships between assemblers and suppliers that were more cooperative than competitive (Addis, 1999).

What arose from this was that the parts sector representing Brazilian capital was supported by the government, and vice versa. Thus, the alliance led to a non-vertical[13] integration of the auto industry. This was to the benefit of parts manufacturers since they could achieve economies of scale by supplying to a large range of assemblers. Under pressure to meet 90–95 per cent by weight domestic content requirements within three and a half years, TNC assemblers relied on local parts manufacturers (Mericle, 1984). The horizontal subcontracting system strengthened the backward linkages of the industry, which were used to legitimate the dominance of foreign capital. GEIA specified that it would encourage non-vertically integrated projects. As stated in the Relatório:

"Despite the knowledge that in industrialised countries, especially in the United States, the automotive industry tends towards a superficial structural organisation, it would not seem reasonable to intend that in its implantation in Brazil, this same industry is going to work based on schemes of vertical integration ... Thus, an excellent opportunity would remain for national investors to operate, possibly on a more economical basis, specialising in the supply of parts and components to various or all aforementioned 'manufacturers' ... recognising therefore the advantage in a horizontal industrial structure, from which emerge two types of producers: manufacturers, primarily foreign, and subcontractors, predominantly national." (Relatório, cited in Shapiro, 1994: 55)

GEIA also thought that an independent parts sector would be more efficient because of the specialisation and the economies of scale attainable by supplying more than one manufacturer. Technological efficiency, therefore, would generate the preferred industrial structure – rigid policies were not required. By 1956, in this favourable climate, there were some 700 parts firms in Brazil, most of them locally owned, employing 90,000 people (da Cruz and da Silva, 1982: 4, cited in Jenkins, 1995: 631).

Finally, international economic circumstances were a factor that influenced technology policy. The success of attracting foreign direct investment (FDI) was due to the nature of international oligopolistic competition in the industry and the promising profits deriving from a large potential market. As the post-war boom came to a close, profits on domestic operations fell in both the United States and Europe, pushing firms to place greater emphasis on overseas expansion. International competition intensified as a result, and aggressive European firms began to challenge the dominant United States firms (GM, Ford, and Chrysler) in their traditional markets, including Latin America. Thus, the timing of the development of the automotive industry synchronised with increased competition between European and American companies for foreign markets. Therefore, Brazil's restrictions upon auto imports led to the anticipated response from the foreign companies to enter the market.

Manufacturing production took place from 1952 to 1962. Initially Ford and GM were reluctant to undertake manufacturing in Brazil. However, by the late 1950s it had become apparent that failure to undertake manufacturing in such a potentially growing market would lead the government to carry out its plan with the aid of other TNCs. When one firm took the first step by deciding to invest in Brazil, other firms followed in order to defend their potential market shares. Under these circumstances, the motor industry in Brazil developed with the strong participation of European capital, through both direct investment and licensing agreements. The goal of import substitution of foreign cars was accomplished in this initial period due to extensive state involvement, pressure from the parts sector, and the international oligopolistic structure of the industry. The results were impressive:

(i) Brazil was one of the first developing countries to succeed in attracting TNCs to undertake automotive plants in the country.[14]
(ii) In addition, it succeeded in creating linkages in the local economy by imposing strict local requirements on assembly companies.
(iii) Between 1952 and 1962, the volume of vehicle production increased by an annual rate of 39 per cent, compared with 10 per cent for manufacturing as a whole (Shapiro, 1994).

4.1.2 The restructuring of the auto industry (1962-1967)

The form of technology policy changed in the mid-1960s due to emerging structural constraints of the industry and politico-economic problems. The most crucial problem encountered after the initial investment phase was product overcapacity: "Firms had built ahead of demand" (Shapiro, 1994: 189). Initially, firms faced high demand. However, once this demand had been met, future sales depended on replacement as well as new demand based on income growth. Demand did not grow as anticipated, due to the high cost of the cars and credit constraints (imperfect capital markets). The initial large investments in capacity by assemblers may also have been part of a strategy of participating firms to gain market share and capture economies of scale, thereby excluding later entrants. Furthermore, according to Mericle (1984), high production costs can be explained by the fragmentation[15] of the Brazilian market, which was divided among too many assemblers. In addition, the high cost of Brazilian parts as well as high tax rates contributed to high sales prices.

More profound reasons for these problems emerge when I bring my political economy framework to bear on the situation, particularly through analysis of the changing relations amongst the state, the part manufacturers, and the TNCs. I also show how this state of affairs, in turn, affected the technology policies adopted to improve the technological performance of the industry.

First, it is important to consider the politico-economic relations between the Brazilian government and local parts manufacturers. The alliance between the state and the supplier firms led to a high percentage of local requirements and to a horizontal integration of the industry. However, this meant that initially the cost of vehicles would be high, which was a major problem for the sales of the cars. In addition, the parts sector was not technologically advanced and could not meet the quality and quantity requirements of assembler firms. Accordingly, the domestic parts sector could not support the assemblers and this was bound to create short-run efficiency problems.

However, soon after firms met the local content requirements, they shifted their interest to in-house manufacture of parts, or managed their own pool of component makers. The vertical structure of the industry made the components sector more fragmented and dependent on large-scale assembler capital. In addition, the monopsonistic power of assemblers led them to control the parts sector and marginalise it (Shapiro, 1994). By the mid-1960s, the tendency towards vertical integration, along with the relative bargaining power of TNCs, appears to have become more pronounced.

Accordingly, politico-economic relations (such as the state's support of parts manufacturers) led to a technology policy that contributed to an increase in the cost of cars. This, in turn, created problems of overcapacity due to low demand. In response to the crisis, the politico-economic relations, which defined the industry, changed. The industry in the following period was characterised by vertical integration and lower local content requirements.

The second issue that should be addressed here refers to the state's inefficiency in imposing entry restrictions on assembly firms. The speed with which car manufacturing was pursued in Brazil led to the uncontrolled entry of foreign firms into the assembly sector. This created great fragmentation in that sector and contributed to high costs. Moreover, the excessive number of entrants led to overcapacity. In spite of the efficient building of institutions, the Brazilian government, in contrast to that of South Korea, lacked the strength to impose entry restrictions on foreign firms.[16] This means that one aspect of the recession had its roots in the failure of the government to use efficient technology policy. Leaving entry decisions to be regulated by market forces yielded negative results. Brazil's failure to impose strict entry regulations derived (and still derives) from the weakness of its bargaining position in relation to that of the TNCs and its fear of retaliatory action by excluded firms.

The technological and economic problems of the industry discussed above were part of a generally deteriorating politico-economic situation in the country.[17] Ultimately, polarisation of economic policy (between the right and the left) facilitated a military coup in 1964. The military regime made a clear choice between efficiency and protecting Brazilian capital. It encouraged consolidation and mergers in order to eliminate excess capacity and rationalise the industry. Brazilian capital was sacrificed in both the assembly and the parts sectors. Moreover, even without state intervention, the crisis led to a greater concentration in the assembly sector of the auto industry,[18] which could achieve higher profits at the expense of reduced wages.

4.1.3 The way to the "Brazilian Economic Miracle" (1968–1974)

In the 1970s, the situation was modified as a result of the restructuring of capital that accompanied the crisis of the international motor industry. Since the domestic market had stagnated, the only way out of the crisis seemed to be to embark on exports to other markets. The Brazilian government mobilised TNCs by offering incentives to export.

The military regime, in order to overcome problems of overcapacity and the low demand of the domestic market, pursued the following policies:

(i) The regime created effective demand by increasing consumer credit and by pursuing income concentration towards the upper and middle class. Even though this policy raised income inequality, it was beneficial to the motor industry. Income concentration and the mobilisation of consumer credit increased the purchasing power of the upper/middle class, which fostered consumption. This played a major role in the rapid expansion of the motor industry.

(ii) In order to compensate for the loss of the domestic market, the state gave export incentives to firms. This period was characterised by a partial shift of the policy from import substitution to promotion of export expansion. The most aggressive Brazilian export incentive was the "BEFIEX" (Fiscal Incentives for Exports), a programme whereby firms could negotiate a package of incentives with the government, including the ability to import capital goods under free trade conditions in exchange for commitment to export (Fritsch and Franco, 1991). Thus, imports of machinery and parts were liberalised through a reduction in tariffs. In addition, there was a new generation of plants,[19] a more efficient scale of production, and modern production techniques. This was a way of overcoming the problems of balance of payment deficits and high cost of production that characterised the import substitution phase of industrial development. The Brazilian state also reduced the local content requirements from 99 to 85 per cent in 1972.

There is a consensus that the auto industry made a fundamental contribution to the Brazilian economic "miracle". There had been substantial growth of the industry, which produced a high of 905,920 units in 1974 (Anfavea, 2002). During the economic "miracle", the motor vehicle industry grew faster than real GDP. The former grew at a rate of 22 per cent per annum, the latter at a rate of 11.2 per cent. The rise of exports was also impressive, more than tripling within two years (in 1972, exports were US$54,146,000 while in 1974 they reached US$203,769,000; Anfavea, 2002). In addition, it has been argued that the industrial strategy was successful, considering the distributed rents between the transnational auto firms and the Brazilian government, and the linkage effects. Shapiro (1994), comparing subsidies and tax revenues, has shown that the amount of taxes paid by vehicle assemblers to the government more than compensated for the direct subsidies they received. In addition, the motor industry was a high linkage sector, because its effects were dispersed among various sectors and generated new ones to produce its intermediate inputs. The parts sector and the steel industry benefited the most. On the other hand, employment did not increase sufficiently, contrary to expectations: the auto industry was capital intensive and did not create many jobs (Hewitt, Johnson and Wield, 1992).

The underlying reasons behind the impressive expansion of the industry and the successful policies in this period can once again be easily explained with the new analytical framework. The reasons are to be found, firstly, in the capacity of the state to credibly engage and pursue a project; and, secondly, in the changing international environment.

The state's role was crucial, in that it encouraged firms to export. The incentives that the Brazilian state provided stimulated export expansion. As Tyler has stated: "By undertaking a host of measures to stimulate manufactured exports, the government reduced the risk attendant in export activities, as perceived by potential exporters ... Increasingly the Brazilian government committed itself to export expansion, and this commitment in itself had a beneficial effect on exports" (Tyler, 1976: 269).

Regarding the international environment, the shift towards exports was essential for the expansion of the industry because it helped to attain economies of scale. However, the initial motivation of the government was to generate foreign exchange rather than the targeting of technological dynamism as such. Also, government policy was not sufficient to drive firms to invest in export plans. The dynamic appearance of Japan in international markets challenged the global "status quo". Toyota, the most important Japanese car manufacturer, concentrated on developing efficient manufacturing processes that could be modified quickly to meet new trends and on building reliable cars. High quality combined with low costs led Toyota to become the fourth biggest car company in the world behind General Motors, Ford, and Chrysler by 1970 (CRS Report for Congress, 1996).

In response, American and European firms were pushed to restructure their operations in order to cut costs and reach Japanese productivity levels. The automotive industry had become increasingly global and competitive in response to a new process of capital internationalisation.[20] The globalisation of the industry becomes evident in the homogenisation of production techniques, the upgrading of performance standards among the three global centres of the motor industry,[21] and the creation of strategic alliances between firms of different national origins. These new circumstances underlined the changing behaviour of firms in their strategies as well as their investment in foreign markets. According to Shapiro: "With increased globalisation, firms became less willing to make investments only to serve an internal market, and foreign subsidiaries potentially became part of a global strategy, no longer serving only individual unconnected markets" (Shapiro, 1994: 224).

Therefore, it becomes apparent that the Brazilian government followed an export strategy that was consistent with the emerging trends in the world auto markets and fitted in with TNC strategies at the time. However, these trends alone are not fully responsible for successful Brazilian export growth. Active state intervention in the form of fiscal incentives also played its part by making exporting more attractive.

4.2 Decline of state involvement in technology policy (1980–2000)

4.2.1 The "lost decade" (1980–1990)

However, changing international conditions made it difficult to maintain the high miracle growth rates. The oil shock in 1974 impacted heavily on Brazil, which imported 80 per cent of its oil. Foreign borrowing and FDI financed the growing trade deficit. Moreover, the oil crisis was one of the main reasons that led the automotive industry into recession. The growth of the motor industry was the prominent factor in the deterioration of the balance of payments. Capital and intermediate goods (machinery, equipment) as well as raw materials were imported to support the needs of growing vehicle production. In addition, repatriated income on foreign loans and direct investment were the main components in increasing pressure on the service account. Sixty per cent of Brazil's increasingly expensive petroleum imports were consumed by motor vehicles (Shapiro, 1994). The negative international environment, along with the reliance of the assembly sector on excessive imports, drove the automotive industry into a deep recession.

The government had to intervene once more in order to prevent a balance of payments crisis. One of the most important government intentions was to limit the expansion of the domestic market[22] of automobiles, in order to reduce oil imports and control inflation. At the same time, the government pursued export promotion by subsidising the auto industry in order to offset the decline of the domestic market. For the transnational subsidiaries, which had invested heavily in the domestic market with the expectation of continued rapid growth, exports served as a temporary outlet for excess capacity. However, this time exports replaced, rather than complemented, domestic sales.

Since the 1980s, there has been a steady decline in technology policymaking. The authoritarian regime placed emphasis upon the macroeconomic stabilisation of the economy at the expense of technology policy. The Brazilian government "forced a swing in the automobile industry's negative balance" by withdrawing all subsidies other than those under BEFIEX (Fritsch and Franco, 1991: 115). During this decade, production fluctuated from 800,000 to 1,000,000 units. This changed after 1992 when production started to increase (Anfavea, 2002).

4.2.2 New perspectives on growth in the auto industry (1990–2000)

Since the end of the 1980s, the influence of the "Washington Consensus" has established a new form of technology policy in Brazil. A partial opening of the economy and a gradual withdrawal of the state from active technology policy characterise the present period. Tariff barriers have been reduced in certain sectors while others remain highly protected. However, in the motor industry tariffs have remained higher than in other industries and are being reduced only gradually. In 1996, the tariffs for companies with plants in Brazil were at 35 per cent while for those without plants they were at 70 per cent. These tariffs were expected to be reduced to 20 and 35 per cent, respectively, by 1999 (Posthuma, 1997: 405; cited in Rodríguez-Pose and Arbix, 2001: 138). In contrast, the car component parts sector was liberalised. Generally, technology policy is more focused and the emphasis has shifted towards enhancing technological capability to produce efficiently and at high quality. The creation or maintenance of a competitive environment and increases in productivity are given equal emphasis in achieving these aims.

An intensive restructuring process in the global auto industry is an additional characteristic of the new period. Companies are integrating their global operations in an attempt to eliminate duplication and cut costs. Moreover, firms, imitating the Japanese model, are reducing the number of parts suppliers they work with, in an attempt to collaborate more closely on product design and development.

During this period, the industry's technological and economic performance has improved significantly. After the lost decade of the 1980s, production and exports have increased rapidly. The recent recovery of Brazil's auto industry coincides with the reduction of trade barriers and deregulation, leading many to conclude that market liberalisation is driving the boom. However, a further examination of the main factors that shape this new picture of the auto industry will reveal the hidden actors and factors that still influence the rules of the game.

While selective reforms have been critical, the revival of the auto industry is not the result of free trade and a non-interventionist state. In fact, despite the overall liberalising trend since 1990, the Brazilian state has developed specific sectoral policies aimed at promoting the industry's recovery. Despite orthodox policy, firms must still satisfy high domestic content requirements (70 per cent). The most important events that influence the new shape of technology policy are as follows:

(i) The Southern Common Market (Mercosur) Agreement in 1994

The Southern Common Market (Mercosur) Agreement is a treaty establishing a Common Market among the Federal Republic of Brazil, the Argentine Republic, the Republic of Paraguay and the Eastern Republic of Uruguay. Greater regional economic integration has taken place with significant potential for all economies to accelerate their processes of economic development. This is a trade liberalisation programme, which consists of progressive tariff reductions accompanied by the elimination of non-tariff restrictions or equivalent measures, as well as any other restrictions on trade between the State Parties. The ultimate aim is to arrive at a zero tariff and no non-tariff restrictions for the entire tariff area (OAS, 2001), but this has not been achieved so far. The Mercosur treaty presents strong evidence of Brazil's efforts to manage trade through industry-specific trade agreements. The agreement facilitates the use of Brazil as an export platform for cars for the region. Moreover, a large market without tariffs and few restrictions in trade is an ideal environment for profit making. Therefore, since 1994 foreign investment[23] has increased significantly in Brazil, especially in the already developed automotive sector.[24] This FDI directly and indirectly influences the shape of technology policy, since foreign players show their interest and try to influence technology policy in their favour.

(ii) The New Automotive Regime[25] (Novo Regime Automotivo)

The New Automotive Regime is a special regime for the auto industry which was established in 1995 within the framework of the 1994 Real Plan.[26] The role of the New Automotive Regime was to consolidate and foster FDI in the automotive sector. Its main objectives were to: (i) keep the large manufacturing plants and the large spare parts companies already installed in the country; (ii) try to restructure existing Brazilian companies; (iii) attract new companies and stimulate the construction of new car plants; and (iv) try to consolidate Mercosur and reinforce Brazil's position as its key economic player (Rodríguez-Pose and Arbix, 2001). Considering the above, even though the shift towards a more liberal strategy by Brazil would logically imply the abandonment of major sectoral policies (as these are inconsistent with neoclassical principles), it is evident that the New Automotive Regime still intervenes heavily in industrial reality.

(iii) Decentralisation and increased emphasis on state formation of technology policy

Since 1980, democratic governments in Brazil have favoured decentralisation and have given more political power to state governors and local mayors. Therefore, individual states have replaced the federal state in drawing up technology policy. Brazilian states, being more independent, offer attractive fiscal as well as locational incentives to attract foreign investors. The issue of fiscal incentives has led to a full-blown tax war among Brazilian states in their attempt to attract more and more FDI in the auto sector.

This is the new reality of technology policy. Technology policy, despite the general climate of liberalisation and minimal state intervention, is still influenced by different players. The role of the state has de facto increased, because of decentralisation. Thus, states formulate technology policy in order to attract FDI and receive a greater share of available benefits. Nevertheless, the federal state also still intervenes and forms sectoral policies, as the "New Automotive Regime" demonstrates. Finally, the elimination of trade barriers between the countries involved in the Mercosur treaty promises greater profits and attracts foreign investors.

Consequently, it has been shown that technology policy is not left to market forces and neither is it governed by a perfect and monolithic state. Rather, Brazil's auto industry has been largely shaped by dynamic interaction between many players, mainly the federal government, state governments, the global auto industry, and also to some extent Mercosur as an emerging entity in its own right.

4.2.3 *Assessment of the industry's technological dynamism*

Since 1995, there has been a tremendous increase in investments in the automotive industry. For example, in 1992 investments in the sector were US$945 million, while in 1998 investment reached US$2,454 million (Anfavea, 2002). Moreover, a large proportion of investments goes to the establishment of new plants: 22 of the 53 industrial automotive units in Brazil (40 per cent) were established in the period 1996–2002 (Anfavea, 2002). As figure 5.1 reveals, production increased rapidly within five years. From 1991 to 1997, production more than doubled, reaching 2,069,703 units for the first time. Exports doubled as well in the same period: in 1997, 416,872 vehicle units were exported. The average rate of export growth in the period 1990–2001 was no less than 9.03 per cent.

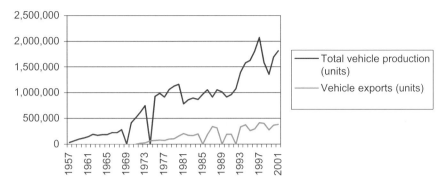

Figure 5.1 Total vehicle production and exports of the automotive industry.
Source: Anfavea (2002).

The above data illustrate that the adopted technology policies ultimately bore fruit. In particular, the fast export expansion suggests that the former infant industry has matured to the extent that it is increasingly able to compete in foreign markets. However, a full assessment of the policy impact cannot be based solely on production and export activities.

The growth of the industry relied on excessive penetration of foreign investors. This raises questions regarding the impact of FDI on the industry as well as on the wider economy. Studies on FDI have shown that there is potential for both costs and benefits deriving from FDI; however, it is not clear yet which side outweighs the other in this case. In the case of the Brazilian automobile industry, the amount of FDI has increased in recent years, but it has also been argued that the expansion of FDI is producing perverse effects in the form of territorial competition among Brazilian states to attract increasingly footloose companies (Rodríguez-Pose and Arbix, 2001). These proactive "development strategies", by both the federal and the state governments, led to an aggressive dispute among states and municipalities, with negative results for the public sector and the economy at large.

Another (somewhat related) aspect of the surge of foreign investments is the destruction of linkages in the domestic economy. The openness of the market, along with the provision of infrastructure (rail, roads, canals, and private port terminals) by competitive states, facilitates the importation of components and parts as well as acquisitions by foreign TNCs of formerly independent Brazilian-owned component makers. The fear is that acquisitions reduce the domestic firms to dependent production plants and lead to erosion of domestic R&D activity and capability. Rodríguez-Pose and Arbix (2001) have argued that this is likely to work against the emergence of local suppliers and, moreover, that it will put

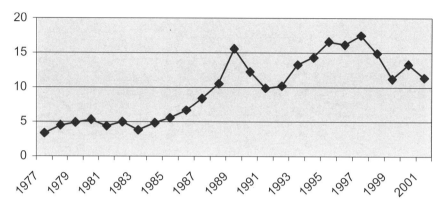

Figure 5.2 Total revenue of auto parts manufacturers.
Source: Anfavea (2002).

the Brazilian component parts industry in a difficult position. As the data in figure 5.2 reveal, the revenues of the auto parts manufacturers increased rapidly until 1989. Afterwards, revenues slowed down as a result of liberalisation and the competition from imports.

Despite this gloomy picture, there is also evidence of increased ability of the auto parts industry to export and compete in international markets. In 1977 the auto parts industry derived 72.8 per cent of its total revenue from supplying the automotive industry, while in 2001 this had reduced to 56.0 per cent (fig. 5.3). This trend demonstrates the strengthening capabilities of the auto parts industry.

5 Conclusions

The new political economy framework presented in this paper sought to investigate the forces that influence technology policy. The framework also provides a basis for understanding and assessing the results technology policy has on the technological dynamism of emerging industries in developing countries, in a more thorough manner than is possible on the basis of the extant monodisciplinary approaches that have been adopted to analyse technology policy. The framework has been applied to the case of the Brazilian automotive industry. The case study illustrates the benefits of the new framework, as follows:
(i) The first period of the industry's development (1952–1980)

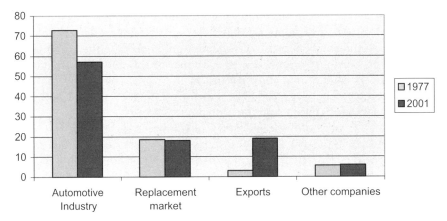

Figure 5.3 Auto parts industry output: Percentage distribution by destination.
Source: Anfavea (2002).

The first period of the industry's development (1952–1980), which was characterised by strong state intervention, has been assessed by many scholars as a successful case of import substitution of cars. The contribution of the auto industry to the Brazilian economic "miracle" is widely accepted. Moreover, the auto industry has strengthened the technological capabilities of auto part/component manufacturers as well as the input supplying industry. However, these externalities are not easy to quantify. They have involved technical assistance, technology transfer, and labour training. It has been demonstrated that the state has constantly intervened to set industrial priorities and improve the competitive position of the industry. In other words, the state has continuously created comparative advantages. The success of the Brazilian auto industry was partly due to the institutional capacity of the Brazilian state to attract FDI, to increase local content (thus creating linkages), and to provide export incentives to TNCs. Therefore, the generation and early technological development of the Brazilian auto industry was believed to be a successful case of state intervention.

Nevertheless, application of the new analytical framework has emphasised that the success of the auto industry cannot be ascribed to the state alone. The dynamic relations among politico-economic agents, arising out of the industrial structure, have influenced technology policy as well. In particular, the influence of the parts and components manufacturers upon the state's decision to establish a motor industry was fundamental. In addition, the oligopolistic competition among TNCs in foreign markets played an important role in determining the entrance of firms into the Brazilian market. Thus, technology policy was not implemented only to overcome market failures.[27] Neither was technology policy successful solely because of state autonomy, despite the creation of "pockets of efficiency" (such as institutions established by the state). Rather, the Brazilian auto industry has been largely shaped by the dynamic interaction between the state (variously democratic and authoritarian), local industrialists, and the global auto industry.

(ii) The second period of the industry's development (1980–2000)

The second period (1980–2000) (as far as can be assessed without sufficient historical perspective) is characterised by a revival of the auto industry. Under the guidance of the "Washington Consensus", the Brazilian economy has become more open and sectoral/industrial policies are not a priority on the agenda of policy makers. This recent recovery of Brazil's auto industry coincides with the reduction of trade barriers and deregulation, leading many to conclude that market liberalisation is driving the boom. However, the "Automotive Regime", along with the increasing competition among the Brazilian states to attract FDI by offering attractive packages of fiscal and other incentives, shows that technology policy still plays a fundamental role in the Brazilian automotive industry. Therefore, despite the announcements of a liberalised economy, technology policy still plays a crucial role and determines the survival and success of the industry.

Existing approaches to the analysis of technology policy were shown to have some weaknesses. Ultimately, in order to understand the effectiveness of technology policy, it is important not only to justify the existence of policies and examine the political conditions that make state intervention effective, but also to look more broadly into the whole range of actors and factors that shape the formation of technology policy. Technology policy *itself* should be an object of analysis and thus become explicit. In my view, therefore, the process of generation, formation and reshaping of technology policy throughout time plays an important role in its ultimate success.

The Brazilian automotive case has shown that this process has not been simple. The dynamic interaction between the state (itself consisting of multiple actors), local industrialists, and the global auto industry played an important role in shaping technology policy, which in turn had a major impact on the technological and economic performance of the Brazilian auto industry. Assuming that technology policy is formed within the higher echelons of a single-actor government and then applied as such is misleading and leads to inability to grasp the essentials of policy failure and success.

Going beyond the debate of whether (or when) technology policy is necessary or not, I reinterpret the historical evolution of technology policy from another stance. The study supports the view that economic, political, and institutional conditions influence the design and implementation of technology policy and thus those conditions should be made explicit in order to understand the process and impact of technology policy. Governments of developing countries as well as aid organisations should view technology policy in its context and take into account the particular characteristics of a country, such as its history, culture, and institutional and social structure.

Notes

1. Technology policy is broadly defined in this paper. In addition to direct innovation-promoting instruments such as R&D subsidies, property rights regulation, etc., it also encompasses policies to promote industrial development more broadly, for example by means of tariffs, quotas, industrial licences, investment subsidies, and taxes. These broader industrial policies are often referred to as "implicit" technology policies. They are often more effective in promoting technological development in developing countries than technology policies in the narrow sense, as the dominant innovation activity in these countries consists of absorption of new technologies generated elsewhere, rather than original R&D.
2. Not all economists share this view. Neoclassical writers argue that industrial development is best left to the market, on the grounds that market-based allocation of resources is most efficient. This fundamental debate in economics will not be dealt with in this paper.
3. A developmental state is one that plays a fundamental role in industrial transformation: the state is essentially a centralised body interacting with the private sector from a position of pre-eminence so as to pursue development objectives (see, for example, White, 1988).
4. In response, the infant industry is expected to exhibit faster productivity growth than the mature industry in developed countries.
5. In older structuralist literature, barriers arising from small markets in low-income countries are also emphasised. These prevent the emergence of industries exhibiting increasing returns to scale. This is known as the static economies argument for infant industries protection (see, for example, Pack and Westphal, 1986).
6. Thailand inherited a complex network of rural roads that were developed by the United States for security purposes. This in turn helped in the agricultural diversification that played an important role in supporting industrialisation (Stein, 1994).
7. After the war, demand for industrial and consumer goods was growing as well as the imports for these items. A big portion of imports consisted of automobiles, contributing to a deterioration in the balance of payments.
8. Local ventures such as Gurgel could not be competitive owing to imperfections in the capital market, and thus they closed down.
9. The fiscal incentives included total exemption of imported machinery and equipment from duties and sales tax, total exemption of complementary parts (not produced in Brazil) from duties, and special access to short- and long-run credit (Mericle, 1984).
10. The foreign exchange advantages had two key provisions: government guarantees of the availability of foreign exchange for the importation of parts (working capital could be raised in Brazil and converted into hard currency for crucial parts imports in the instal-

lation phase); and government subsidies of foreign equipment imports by guaranteeing the availability of hard currency at favourable exchange rates (Mericle, 1984).
11. The major firms involved include Mercedes-Benz, Fiat, Ford, Saab-Scania, Volkswagen, and General Motors.
12. The Law of Similars included various types of protection devices for goods produced locally which were similar in nature to the imported goods. It consisted of the following elements: (i) prohibition of importation of parts that were also produced in Brazil, thus guaranteeing a market for the Brazilian producers. Import licences were issued only for those parts not produced domestically; (ii) prohibition of the importation of non-essential assembled motor vehicles; and (iii) prohibition of importation of completely knocked down vehicles as self-contained units, and a requirement to use any individual parts available domestically (Mericle, 1984).
13. Non-vertical or horizontal integration of the industry means that a component manufacturer could supply its specialised product to every assembler.
14. In the mid-1950s, Mercedes-Benz, Volkswagen, Ford, Scania, and Toyota opened assembly plants in the ABC area (this is the greater area of São Paulo).
15. While the number of firms was too great to take maximum advantage of scale economies, the Brazilian industry was not so excessively fragmented as those of some other countries. For example, "in 1964 in the free port of Arica in Chile there were 20 firms producing 25 models, the largest of which produced 1,500 units" (American Embassy, Santiago; cited in Jonson, 1967: 205).
16. South Korea used tough entry regulations in order to prevent fragmentation of the final assembly sector. It has been argued that relaxation of entry regulations in 1977 in South Korea led to excess capacity and excessive competition (Chang, 1998).
17. In 1963 the rate of inflation reached 78 per cent, gross domestic product fell by 1.3 per cent and foreign investments also fell. In addition, real wages decreased (Mericle, 1984).
18. By this time the assembly industry, mainly controlled by foreign firms, had shrunk from eleven to eight firms. Volkswagen, GM and Ford controlled 89 per cent of the total vehicle market.
19. In the 1970s, new vehicle factories opened outside the São Paulo ABC area: Volkswagen and General Motors installed plants in the Paraíba Valley, Volvo opened a truck and bus plant in Paraná, and Fiat invested in Minas Gerais.
20. According to Jenkins (1987), internationalisation of capital was a necessary strategy in order to overcome the crisis created by the Japanese penetration in Western European and United States markets. This generated the need for a restructuring of capital in order to find new ways of accumulation.
21. North America, Western Europe, and Japan.
22. This policy was pursued by limited consumer credit and by imposing import controls.
23. Foreign companies announced investments in the period 1995–2000 of US$17,244 billion (CEPAL, 1997).
24. In the manufacturing industry 41 per cent of FDI goes into the automotive and replacement market (CEPAL, 1997).
25. This is a programme of investments and exports with a special regime for importation. An industrial enterprise established in a chosen country (newcomer) that commits to invest and export will as a reward be authorised to import capital goods and vehicles with reductions in import tax.
26. The Real Plan was a stabilisation programme established in 1994 by Fernando Henrique Cardoso. Its main objectives were to: (i) reduce inflation; (ii) achieve long-term sustainable growth in output, investment, employment, and productivity; and (iii) achieve a steady and substantial reduction in social imbalance (Freire de Oliveira Santos, 1998).
27. Market failures were plentiful in the Brazilian economy. Imperfect capital markets and the absence of a mature industrial class were the main constraints.

REFERENCES

Addis, C. (1999) *Taking the Wheel: Auto Parts Firms and the Political Economy of Industrialization in Brazil*, University Park, Penn.: The Pennsylvania State University Press.
Amsden, A. H. (1989) *Asia's Next Giant – South Korea and Late Industrialization*, New York: Oxford University Press.
—— (1997) "Bringing Production Back In – Understanding Government's Economic Role in Late Industrialization", *World Development* 25(4): 469–480.
—— (2001) *The Rise of "the Rest": Challenges to the West from Late-industrializing Economies*, Oxford: Oxford University Press.
Anfavea (Associação Nacional dos Fabricantes de Veículos Automotores; National Association of the Manufacturers of Automotive Vehicles) (2002) "Anuario Estatístico da Indústria Automobilística Brasileira" (Statistical Yearbook of the Brazilian Automotive Industry), online at: http://www.anfavea.com.br/Index.html.
CEPAL (1997) "La Inversión Extranjera en América Latina y el Caribe: Informe de la CEPAL" (Foreign Investment in Latin America and the Caribbean: CEPAL Information), Santiago de Chile: United Nations, online at: http://www.eclac.cl/publicaciones/desarrolloproductivo/5/lcg1985/capiid.html (p. 45).
Chang, H. J. (1994) *The Political Economy of Technology Policy*, London: Macmillan Press.
—— (1998) "Korea: The Misunderstood Crisis", *World Development* 26(8): 1555–1561.
Corden, M. (1974) *Trade Policy and Economic Welfare*, Oxford: Clarendon Press.
CRS Report for Congress (1996) online at: http://cnie.org/NLE/CRSreports/international/inter-69.cfm.
Dicken, P. (1998) *Global Shift. Transforming the World Economy*, London: Paul Chapman.
Evans, P. (1995) *Embedded Autonomy: States and Industrial Transformation*, Princeton: Princeton University Press.
Freire de Oliveira Santos, F. (1998) "Fiscal Adjustment in the State of Bahia", final paper presented to the Institute of Brazilian Business and Public Management Issues, as part of the Minerva Program, Washington, D.C., Spring 1998.
Fritsch, W. and G. H. B. Franco (1991) *Foreign Direct Investment in Brazil – Its Impact on Industrial Restructuring*, Paris: OECD Development Centre.
Gordon, L. and E. L. Grommers (1962) *United States Manufacturing Investment in Brazil: The Impact of Brazilian Government Policies, 1946–1960*, Boston, Mass.: Harvard University Press.
Hewitt, T., H. Johnson and D. Wield (1992) *Industrialization and Development*, Oxford: Oxford University Press.
Jenkins, R. (1987) *Transnational Corporations and the Latin American Automobile Industry*, London: Macmillan Press.
—— (1995) "The Political Economy of Technology Policy: Automobile Manufacture in the Newly Industrialising Countries", *Cambridge Journal of Economics* 19(5): 625–645.

Jonson, J. L. (1967) "Problems of Import Substitution: The Chile Automobile Manufacture Industry", *Economic Development and Cultural Change* 15 (January): 202–216.

Lall, S. (1992) "Technological Capabilities and Industrialization", *World Development* 20(2): 165–186.

Martinussen, J. (1999) *Society, State and Market. A Guide to Competing Theories of Development*, London: Zed Books.

Menzies, H. (1997) *Canada in the Global Village*, Ottawa: Carleton University Press.

Mericle, S. K. (1984) "The Political Economy of the Brazilian Motor Vehicle Industry", in R. Kronish and K. S. Mericle, eds., *The Political Economy of the Latin American Motor Vehicle Industry*, Cambridge, Mass. and London: The Massachusetts Institute of Industrial Press, pp. 1–40.

OAS (Organization of American States) (2001) "Southern Common Market (MERCOSUR) Agreement", Sistema de Información al Comercio Exterior (SICE –Foreign Trade Information System), online at: http://www.sice.oas.org/trade/mrcsr/mrcsrtoc.asp (accessed on 24 August 2003).

Oman, C. (1994) *Globalisation and Regionalisation: The Challenges for Developing Countries*, Paris: OECD.

Pack, H. and L. E. Westphal (1986) "Industrial Strategy and Technological Change: Theory versus Reality", *Journal of Development Economics* 22(1): 87–128.

Rodríguez-Pose, A. and Arbix, G. (2001) "Strategies of Waste: Bidding Wars in the Brazilian Automobile Sector", *International Journal of Urban and Regional Research* 25(1): 134–154.

Shapiro, H. (1994) *Engines of Growth: The State and Transnational Auto Companies in Brazil*, Cambridge: Cambridge University Press.

Stein, H. (1994) "The World Bank and the Application of Asian Technology Policy to Africa: Theoretical Considerations", *Journal of International Development* 6(3): 287–305.

Tyler, W. G. (1976) *Manufactured Export Expansion and Industrialization in Brazil*, Tübingen: Mohr.

UNCTAD (2000) *The Competitiveness Challenge: Transnational Corporations and Industrial Restructuring in Developing Countries*, Geneva: United Nations.

Wade, R. (1990) *Governing the Market: Economic Theory and the Role of Government in East Asian Industrialization*, Princeton: Princeton University Press.

White, G., ed. (1988) *Developmental States in East Asia*, London: Macmillan.

6
Technological learning in small-enterprise clusters: Conceptual framework and policy implications

Marjolein C. J. Caniëls and Henny Romijn

1 Introduction

In many parts of the developing world, small and medium industrial enterprises (SMEs) are currently being confronted with formidable competitive challenges. Few are still able to insulate themselves from the pervasive effects that economic liberalisation and deregulation are having on their local economies, particularly through international trade and foreign direct investment. For some firms, new business opportunities emerge as international production and trade chains extend into far-flung places in continuous search of cheap sources of supply. Many others are challenged to defend their traditional home markets – the mainstay of the great majority of small firms in less developed economies – against new dynamic competitors.

Recent research has begun to throw light on how SMEs are responding to these processes, and to identify suitable types of assistance. One salient finding is that their competitiveness could be boosted by being part of regional industrial clusters – agglomerations of actors engaged in similar and complementary activities. A cluster may be defined as a group of business enterprises and non-business organisations that share a common regional location where the region is defined as a metropolitan area, labour market, or other functional economic unit.[1] There is by now substantial evidence that SME clusters are widely prevalent in many industrial sectors in a range of developing countries, from low-income countries such as Ghana and Kenya in Africa to middle- and higher-income countries in Asia and Latin America.[2]

Of course, clustering is neither a necessary nor a sufficient condition for industrial dynamism. Many clusters in less developed countries have existed for a long time without showing substantial signs of innovation or growth. Conversely, there are numerous cases of innovative non-clustered firms. However, it has been noted in the recent literature that clusters do have features that may facilitate dynamic growth. In particular, it is claimed that SME clusters may be able to boost regional development by creating possibilities for accumulating capital and skills through "collective efficiency" (Schmitz, 1995; Schmitz and Nadvi, 1999).

However, so far cluster research has concentrated mainly on the economic benefits to which clusters may give rise, while the technological factors that supposedly underpin these benefits have been given rather cursory treatment. The main analytical focus of studies in this line is on the meso level, while the internal functioning of the individual firms that make up a cluster remains largely a "black box" (Bell and Albu, 1999). Some writers have drawn attention to factors such as quality upgrading (Nadvi, 1999a,b) and incremental innovation (Mytelka and Farinelli, 2000), but the study of these issues has not been based on a micro-economic theoretical foundation of firm behaviour and strategy.

Yet, development of technological competence by small firms is arguably crucial for them to be able to hold their own in fast-changing and fiercely competitive markets. Flexible and quick adjustment and adaptation require technical skills, knowledge, and organisational/managerial capacity – technological capability in short – to make the right investment choices, increase productive efficiency, meet tight deadlines, and engage in continuous upgrading of quality and design (Bell and Albu, 1999; Romijn, 2001). Since only a very small minority of SMEs in developing countries are already well equipped for these tasks, we posit that regional industrial growth and competitiveness in these countries ultimately cannot be sustained without advancement of the technological capability base of individual firms. Likewise, effective policies aimed at promoting competitiveness of regional clusters cannot be based solely on insights about cluster dynamics. In addition, they need to build on knowledge about the internal technological functioning of the companies operating in the clusters, and they have to address the main bottlenecks that occur at that level.[3]

The aim of this paper is to contribute to the current policy debate about promoting regional economic growth and SME competitiveness in development by opening up the black box of the firm by using the collective efficiency approach. We do this by linking this approach to the so-called "technological capability" literature, which sheds light on the micro-level technological underpinnings of industrial competitiveness in a developmental context. The unified conceptual framework that results from joining the meso and micro approaches is then used to shed new light on the forces that determine the competitiveness and growth of industrial SME clusters in development, and to derive new guidelines for policy.

The new conceptual framework is introduced in section 2. In section 3 we spell out the main new policy insights that result from adopting the framework, showing its added value by contrasting the policy insights with the sort of policy implications that are typically derived by the two partial approaches on their own. Section 4 contains an empirical case study of farm equipment manufacturing in Pakistan's Punjab Province, which serves to illustrate the additional value of the new approach for policy design in practice. Section 5 presents the conclusions.

2 Joining the meso and micro approaches

The meso-level collective efficiency (CE) approach and the micro-level technological capability (TC) approach yield complementary insights into the determinants of the long-run competitiveness of SMEs. CE studies analyse the effects of clustering on the economic performance and competitiveness of clusters and countries, whereas the TC approach studies how intra-firm learning processes affect long-term competitiveness.

Figure 6.1 shows how we intend to join the two approaches. The analytical perspective of the CE approach is represented in the left-hand side of the figure. Studies in this area put strong emphasis on the advantages for firms that accrue from being part of an industrial cluster. These gains tend to boost the economic performance of the industrial cluster and therefore the region as a whole. Many studies in this tradition have identified a range of advantages that are induced by clustering, particularly through possibilities for easy networking and collaboration.[4] The most recent analytical contributions in this line of research have also drawn attention to the positioning of clustered firms in the value chain, and how the governance of these chains affects the possibilities for local upgrading (Humphrey and Schmitz, 2000, 2001; Nadvi and Halder, 2002). This is especially important in global value chains in labour-intensive consumer goods sectors, where access to new information and knowledge about technology and markets by firms in local clusters is dependent on dominant actors such as importer-retailers and importer-wholesalers (Schmitz and Knorringa, 2000). However, as stated earlier, the effects of clustering on knowledge accumulation processes in individual firms have not been studied in depth.

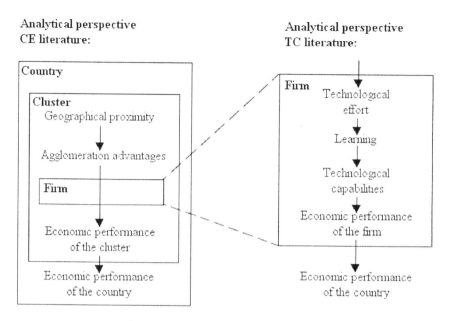

Figure 6.1 Integrating macro with meso

The right-hand side of figure 6.1 presents the analytical perspective taken in the TC literature, which we shall use to open this black box. In contrast to the CE literature, the TC literature puts intra-firm knowledge accumulation processes centre stage. The term "technological capability" was first coined in the early 1980s by researchers probing intra-firm technological dynamics far from the world's technological frontier.[5] Inspired by upcoming evolutionary theories of technological change (later culminating in Nelson and Winter, 1982; Dosi, 1988), they showed importation of new technologies by itself to be insufficient for enhancing productivity and inducing self-sustaining industrialisation. Mere access to foreign technology – whether in the form of plant and machinery or documentation and blueprints – does not imply mastery over it (Dahlman and Westphal, 1981). The impossibility of fully codifying new knowledge, and the fact that foreign technologies are less than perfectly suited to specific local needs and conditions, can be powerful barriers to the effective implementation of new technologies in a new setting. Therefore, accumulating technological capability requires considerable technological efforts – investment in time and resources aimed at assimilating, adapting, and improving known technologies, and (ultimately) creating new technologies in-house. A lengthy learning process is usually involved. The capabilities that firms acquire help to improve their economic performance, and (by assumption) regional and national economic performance as well. However, studies in this area have not systematically addressed the question of why and how spatial proximity of groups of actors may boost firms' individual learning efforts and/or increase the effectiveness of those efforts. A firm's technological improvement efforts are seen to be somehow pushed and enhanced by various external stimuli (including governance of global value chains), but no special effects of proximity to other local parties have been noted.

Figure 6.1 gives a good representation of the way in which we join the meso-level CE perspective with the micro-level TC perspective. Looking from left to right in figure 6.1 along the dotted lines which connect the left-hand side to the right-hand side, we zoom in on the firm level, which is not explicitly analysed in the CE approach. The crux of the integration of the two approaches then revolves around the question as to how exactly the connection between the meso and the micro level is made. In other words: What are the mechanisms by which the different agglomeration advantages that are commonly observed in an industrial cluster translate into increased and/or more efficient technological effort within individual firms in that cluster? Once this is known, we will be able to understand how clustering can enhance firm-level technological learning, leading to more advanced capabilities and better economic performance, and (ultimately) how it contributes to regional and national economic growth.

We need two pieces of information in order to answer this question.[6] Firstly, on the micro side we need to know the different types of technological efforts commonly undertaken by firms. A useful classification has been developed by Bell, based on a survey of a range of empirical capability studies undertaken in different developing countries. Several important types of efforts were commonly noted, namely staff training, staff hiring, in-house technological improvement (including R&D), and external search for information about new technologies and markets (Bell, 1984).[7]

The second item consists of the main types of agglomeration advantages emanating from geographical clustering by firms. Information about this is provided from the literature on the meso side of figure 6.1. In the CE literature, the common distinction has been between economies emanating from so-called "passive" and "active" collective efficiency. The former occur spontaneously, whereas the latter require active and purposive joint action among cluster participants (Schmitz and Nadvi, 1999). Looking more closely, however, one may notice that both passive and active collective efficiency are composed of the same two categories of agglomeration economies that have been distinguished conventionally by economists (for an overview, see Malmberg and Maskell, 2002). The first of these comprises economies of scale, scope, and transaction, which are all kinds of cost advantages that accrue from firms locating close to each other (Marshall, 1920; Richardson, 1978). The second category consists of technological or knowledge spillovers, which are intellectual gains through exchange of information for which a direct compensation to the producer of the knowledge is not given, or for which less compensation is given than the value of the knowledge (Audretsch and Feldman, 1996; Caniëls, 2000; Feldman and Florida, 1994; Marshall, 1920). We adopt this conventional classification below, but we will return to the passive–active distinction in the policy discussion in section 3.

Since the two main groups of agglomeration economies are still rather heterogeneous, it is useful to make some further subdivisions. Within the category of scale, scope, and transaction economies, we identify two groups according to the two main types of economic activity that firms pursue: (I) economies of scale, scope, and transaction *in the production of goods and services*; and (II) economies of scale, scope, and transaction *in undertaking technological effort*. Further, we distinguish three groups of knowledge spillovers, following a classification adopted by Stewart and Ghani (1991), namely, spillovers emanating from: (III) changing attitudes and motivations; (IV) human capital formation through informal learning-by-doing; and (V) transfer of technological information.

We now proceed to examine the different ways in which these five agglomeration advantages may contribute to the four types of firm-level technological efforts identified by Bell (1984). The discussion is summarised in table 6.1. The five rows in the table represent the types of agglomeration advantages introduced above, while the four columns, labelled A through D, represent the types of technological effort. The contents of the cells describe the (sub-) mechanisms through which the agglomeration advantages affect these technological efforts.[8] Some types of agglomeration advantages, particularly II and V, can influence technological effort in several distinctly different ways. This is shown in the table by means of a finer subdivision within the main rows.

We discuss the table row-wise. Row I indicates a mechanism associated with direct cost advantages in production obtained by clustered firms. One such cost saving emanates from high demand (Swann, 1998). Since clustered firms reap more economies of scale in production compared to non-clustered firms, they are left with more financial resources to invest in technological effort. Another source of cost savings lies in the fact that clustering may induce more intensive competition among input suppliers, which reduces input costs for user firms (Nadvi, 1999b). Both types of cost savings may affect all types of technological effort (columns A through D).

Row II indicates that economies of scale, scope, and transaction in

Table 6.1 Direct effects of agglomeration advantages on the technological efforts (TE) of the firm

Agglomeration advantages	A. Hiring	B. Training	C. Information search	D. R&D
I. Economies of scale, scope, and transaction in production	Lower unit cost due to large market size leaves more resources for technological effort.			
II. Economies of scale, scope, and transaction in knowledge accumulation		a) Large local market gives rise to critical minimum demand for innovations, inducing technological efforts to develop them. b) Presence of specialised suppliers lowers transaction costs, which facilitates easy and cheap access to specialised inputs needed for technological effort.	c) Low transaction costs facilitate joint undertaking of technological efforts, thus leading to cost savings. d) Low transaction costs stimulate *additional* technological effort in joint indivisible and complementary projects, which in turn facilitates access to, and leads to, generation of new information and knowledge.	
III. Knowledge spillovers: changing motivation and attitudes		Exposure/demonstration effect stimulates demand for TE.		
IV. Knowledge spillovers: human capital formation through informal learning-by-doing	a) Exposure/demonstration effect stimulates demand for TE.	b) Direct free input through industry-wide accumulation of skills.		
V. Knowledge spillovers: technology transfer	Direct free input through inter-firm movement of trained labour.		Direct free knowledge input through trade journals, meetings, fairs, etc. Direct free input through user–producer interaction.	

knowledge accumulation itself may have four significant effects on technological effort. Firstly, clusters can generate a critical minimum demand for new, specialised products or services that cannot be produced profitably elsewhere. In turn, this stimulates investment in efforts to master the production of these new items (IIa) (Stewart and Ghani, 1991). This effect may apply to all kinds of technological effort (columns A through D).

A second important link deals with the local presence of suppliers of specialised inputs who are attracted by large local demand. This may lower transaction costs associated with procurement of specialised inputs. Thereby, clusters act to reduce costs of specialised inputs needed to undertake investments in technological effort (IIb) (Tewari, 1999). This mechanism may again influence all kinds of efforts, because there are manifold actors offering specialised services, including workers with specialised skills and technical consultants (A), institutions providing training courses (B), government extension services (C and D), sourcing agents looking for suitable suppliers (C), suppliers of machinery, materials and components (C and D), and so on.

A third important link in row II operates by offering possibilities for firms to join networks of innovators because of low transaction costs associated with local interaction (IIc) (DeBresson and Amesse, 1991; Freeman, 1991). This leads to cost advantages from sharing costs and risks. Existing literature pointing to this mechanism relates primarily to economically advanced countries, so that the focus has been primarily on formal R&D-type efforts (D), but it is no less likely to work with respect to more informal types of effort with scope for collective investment. In economically less developed settings, these activities are more likely to take the form of training (B) and search (C) than R&D.

In another study pertaining to advanced countries (Baptista, 1998), it has been pointed out that pooling R&D resources will induce *more* R&D investment as well, as it becomes feasible to embark on large, costly projects that are beyond the capacity of individual investors (IId). A variation on this theme is the case where proximity allows parties to invest in technological effort that requires mutual commitment, since they need to supply complementary inputs for it. As in the case of row IIc, this mechanism might operate just as well in the form of more informal technological effort in less developed economies. In particular, training (B) and search (C) are two non-R&D-based efforts for which collective investments are also feasible.

Rows III, IV, and V indicate that knowledge spillovers from other firms may complement a firm's own efforts and thereby increase the efficiency of those efforts. Implementing knowledge from outside the firm increases its chances of success. Firms might benefit from complementarity and synergy effects that arise from the technological improvement activities undertaken by other firms in the cluster. Spillovers are facilitated by opportunities for firms to establish direct contact with each other in a cluster, such as through inter-firm labour mobility and formal and informal exchange of information and ideas (Baptista, 1998; Feldman, 1994; Nelson, 1993; Von Hippel, 1988).

Changing attitudes and motivation (III) primarily works by exposing people to new ideas and artefacts in a particular environment (Stewart and Ghani, 1991). These act on people's mental predisposition in such a way that they will begin to favour change over stability, and thereby stimulate investment in the technological efforts needed to bring it about. These advantages affect firms' efforts in a broad manner. For example, changing attitudes happen through exposure to new information, ideas, and products, which generally stimulates demand for technological improvement efforts of all kinds (A through D).

Human capital formation through informal learning-by-doing (IV) likewise acts through changing attitudes, in this case attitudes towards work (IVa) (ibid.). Like mechanism III, it is a broad effort-inducing mechanism (A through D). In addition, learning-by-doing entails assimilation of a basic body of more specific production-related technical knowledge and skills that are common in a local industrial environment (IVb) (ibid.). This constitutes a direct free input complementing a firm's own investments in staff training (B). Thus, this spillover affects not only the demand for technological effort, but also the supply of inputs for it.

Technological transfer (V) acts entirely on the supply side. It operates through three channels: inter-firm movement of trained labour (A); trade journals, meetings, trade fairs, and various other fora for inter-personal exchange (C and D); and user–producer interactions which often occur in the course of implementing and perfecting innovations in iterative fashion (also C and D) (ibid.).[9] Inter-firm movement of trained labour boosts skill levels through hiring of new staff, while communication fora and user–producer interactions are primarily sources of free new information and knowledge about technologies and markets, which complement the firm's own search and research efforts.

Technology transfer spillovers often interact with economies of scale, scope, and transaction. Low transaction costs in clusters directly facilitate (horizontal and vertical) business interaction, joint projects, and labour mobility, which are the main vehicles through which skills, knowledge, and ideas travel across firms. Furthermore, we have seen that economies of scale, scope, and transaction boost the amount of intra-firm technological effort in various ways. Clearly, the more actively firms are engaged in learning, the more spillovers to neighbouring firms are likely to result. The recipients essentially receive free inputs that complement their own technological efforts and in this way increase the effectiveness of their learning processes. In sum, when economies of scale, scope, and transaction work in tandem with knowledge spillovers, both the amount and effectiveness of intra-firm technological effort will receive a boost.

So far, influences from outside the cluster (i.e. value chains and macro-economic conditions in general) have not been explicitly considered in the above discussion, as the main emphasis of this paper is on exploring the effects of *local agglomeration* on firm-level technological dynamism. However, factors external to the cluster could of course also exercise a major impact on firm-level technological dynamism. For instance, technological effort is always an investment decision, and therefore subject to risk and uncertainty. Uncertainty about economic conditions, both locally and globally, will affect effort-related decisions, and thereby economic growth performance in clustered firms (as well as non-clustered ones). Chain governance will also affect technological dynamism of firms in local clusters, by facilitating or obstructing the vertical diffusion of new knowledge and skills about technology and markets. The more local actors are able to tap into knowledge and information from outside their cluster, the more local spillovers are likely to result, in turn facilitating knowledge creation in individual firms.

All mechanisms discussed so far referred to positive effects of agglomeration on technological effort. However, clustering sometimes gives rise to negative effects as well, which counteract the positive forces to some extent. These negative effects have not been listed in table 6.1 in order to avoid excessive complexity. However, they cannot be ignored, particularly when it comes to deriving policy lessons. At least three potentially important negative effects should be mentioned. Excessive competition among small producers who are incapable of differentiating their products substantially from those of others may squeeze margins, leaving fewer resources for technological improvement. Moreover, in clusters where secrecy is hard to maintain and legal protection of innovations is non-existent, knowledge spillovers may also have drawbacks as they reduce innovation incentives for the party that generates them. Further, there could be considerable social barriers to undertaking technological improvement efforts in change-resistant traditional communities. Such negative effects are more likely to dominate positive ones in adverse economic settings, for example in the case of stagnant demand.

3 Policy implications of the integrated approach

Our integrated framework has distinct added value for policy-making, in comparison with the two partial approaches.

As explained above, the main policy prescriptions of the CE literature hinge upon the distinction between passive and active CE. The main contributions to this literature expect a lot from policies aimed at promoting active CE, through strengthening institutions that promote joint action, inter-firm collaboration, and horizontal and vertical networking (Ceglie and Dini, 1999; Stewart and Ghani, 1991; UNCTAD, 1998). Collaboration with buyers is seen to be especially important, as understanding customers' needs helps producers to tackle key competitiveness problems. In contrast, the passive, non-collaborative externalities are essentially taken as given in the CE approach.

The integrated framework in table 6.1 suggests a broader range of policy instruments than that offered by the CE studies. It is of course possible, on the basis of the table, to make a similar distinction between agglomeration advantages that need active collaboration and networking to bring them about (notably IIc and IId), and others that occur without purposive action of this sort. However, there are no a priori reasons why active collaboration should be singled out in our framework as the only feasible object of the interventions. Indeed, many of the spontaneous, non-collaborative mechanisms could also be amenable to further stimulation. Possibilities in this direction appear to exist especially in the sphere of knowledge spillovers. A relevant example would consist of measures to upgrade basic physical infrastructure facilities such as roads and telephone networks, which may indirectly induce more inter-firm spillovers by facilitating trade-based interactions and exchange and by increasing labour mobility through the mechanisms listed in row V of table 6.1. Another example would be the establishment of education and training facilities that help to make people mentally more receptive to the adoption of new production techniques by exposing them to the same. Once a few firms have adopted a new technique, in due course it is bound to be adopted more widely and contribute to the spread of knowledge through demonstration effects and other mechanisms identified in rows III and IV.

The broader policy focus suggested by our framework has particular advantages in those situations where intensive inter-firm cooperation in clusters has proved difficult to get off the ground and/or hard to sustain. This has indeed been the case in many of the clusters examined in the CE studies (see, for example, Nadvi, 1999a,b; Schmitz, 1999; Visser, 1999). In these kinds of environments, measures aimed at facilitating individual knowledge absorption and subsequent spontaneous diffusion among firms may turn out to be a more effective and sustainable policy strategy. Our framework further suggests cost-effective ways of providing some of these types of support, for example in the sphere of education and training. Targeting a small number of progressive companies with programmes to improve work practices, upgrade quality, and so on, may be sufficient. Spillovers will ensure the gradual diffusion of the new knowledge and skills throughout the cluster, through demonstration effects (III and IVa), industry-wide accumulation of skills (IVb), inter-firm movement of labour (V-A), and circulation of information and knowledge (V-C and D).

The framework laid out in table 6.1 also suggests a somewhat more focused approach to intervention compared with the CE approach. Given the central policy goal as stated in the Introduction – namely, the structural improvement of regional competitiveness – interventions need to be directed as closely as possible towards the creation of dynamic learning economies. This is achieved by directing assistance towards stimulating technological effort. The sort of collaboration-focused support advocated in CE studies does not appear to be tightly focused on these dynamic effects. Some of the policy instruments suggested in the CE approach seem to be geared more towards generating static production economies. For example, strengthening collaborative institutions may help producers to lobby for common access to credit, joint purchasing of inputs, and tax relief, but there is no direct link between such activities and technological learning. Moreover, to the extent that the CE studies do address dynamic "upgrading" issues, they lack specific policy guidelines. For example, in their recent summary paper about the CE literature, Schmitz and Nadvi state that, "In more advanced clusters, policy measures have to be geared to promoting upgrading, in particular in areas of technical learning and innovation". A first step is "benchmarking local practices against those of global market leaders" (Schmitz and Nadvi, 1999: 1509–1510). Yet no details are given about the types of learning, innovation, and work practices actually involved and the policy instruments that could be used.

The application of our integrated framework affords more specific policy conclusions in this direction, since it examines firm-level processes of technological improvement. To do this, we used an established analytical approach (TC), which has provided a range of policy lessons that are in principle applicable to clusters as well as to firms individually. Its policy findings rest on the recognition of pervasive market failures in the development process. On the one hand, market-oriented interventions to stimulate demand for innovations are considered essential, as these create incentives for firm-level investment in technological effort. Relevant instruments include trade and industrial policies that regulate market structure and competition; fiscal, legal, and investment policies that regulate access to financial capital and foreign technologies; and general economic policies that promote a stable macro-economic climate and economic growth. These demand-stimulating interventions are complemented by supply-side assistance aimed at supporting firms' learning processes through the provision of critical missing inputs. Important supply-side measures are the creation of an information-rich environment, the building of science and technology support institutions, education, and training (Lall, 1992; UNCTAD, 1996).

However, since the TC approach has not paid specific attention to geographical agglomeration, its policy guidelines do not take account of any special advantages for learning and growth that obtain in clusters. Our integrated approach does make this possible, through the insights gained from the meso literature on the left-hand side of figure 6.1.

Firstly, conventional TC measures are potentially more effective when supplemented by policies to induce firms engaged in similar and complementary activities to co-locate. This is not meant to be a recommendation for a souped-up version of the old industrial estates programmes that were pursued by many developing country governments in the 1960s and 1970s. These have generally been disappointing, among other things owing to the fact that insufficient account was taken of the need to ensure complementarity and similarity in the industrial activities located on the estates. What we propose is that governments should try to facilitate clustering by private sector actors engaged in similar and complementary activities through appropriate zoning regulation and offering attractive location-based infrastructure facilities, as in new technology parks in more advanced countries.

Secondly, the framework provides specific handles for enhancing capability accumulation in firms that are already located in clusters. In particular, several of the learning-enhancing mechanisms identified in table 6.1 can be leveraged through targeted interventions. Relevant policies include, for example, collective institution-building initiatives that promote inter-firm trust. These may lead to cost savings and may induce new large knowledge-building projects that are beyond the scope of individual firms. They may also indirectly trigger various inter-firm knowledge spillovers. Other policies could aim at stimulating spillovers directly, by concentrating new technological and market information, training, and extension services locally in clusters.

Thirdly, the framework suggests once again that such interventions do not need to aim at directly targeting all potential beneficiaries in a locality, as the presence of local spillovers ensures quick diffusion of knowledge and skills (see above). This runs counter to the argument, which has often been advanced among small enterprise development practitioners and researchers, that dualism within the small-scale sector will increase if assistance is exclusively focused on a few advanced participants. This argument may be valid in general, but we expect that it carries less weight in local clusters. Local clusters facilitate trickle-down, even if the local firms do not actively network or collaborate. Less forward-looking firms in a locality may be expected to benefit to some extent from policy interventions that are not directly aimed at them, for example through spontaneous demonstration effects and other spillovers.

The additional value of these policies could be outweighed to some extent by various negative effects that potentially also emanate from clustering (section 2). These effects are likely to play a role of some importance, especially in clusters whose firms are still functioning at a very low technological level. Producers in that situation usually lack the skills to diversify their products substantially from each other, so that keen competition is hard to avoid. Copying is also bound to be rampant in this kind of environment. Furthermore, in traditional communities the threat of social sanctions on deviant behaviour may pose a considerable constraint on investment in technological innovation. It is important that policy makers be aware of these kinds of problems, but they cannot be remedied quickly and easily. Sometimes, measures to foster inter-firm collaboration and trust may help, but in general clusters simply need time to outgrow this stage.

It has been emphasised by cluster researchers in the collective efficiency tradition that the orientation of policies will have to take account of heterogeneity in cluster performance. A distinction is made between survivalist clusters on the one hand, and more advanced and dynamic clusters on the other. In survivalist clusters, the need to foster networking and collaboration is stressed, while policies aimed at advanced clusters would need to be geared more to technological upgrading and innovation (Schmitz and Nadvi, 1999: 1509). While recognising that cluster performance differs widely across countries, regions, and sectors, we do not follow this distinction. In survivalist clusters with high competition, effective inter-firm networking and collaboration are likely to be particularly difficult to establish, perhaps more so than in more advanced clusters. Thus, survivalist clusters are likely to benefit more from policies aimed at creating conditions under which spontaneous spillovers will flourish.

It seems more useful to take account of cluster heterogeneity by tailoring the *content* of the policies specifically to the level of their technological advancement. In that sense, our approach follows the policy directions deduced from the technological capability studies, which have emphasised that technology policies have to be tailored to the economic and technological level of the countries concerned. The same principle is applicable at the cluster level. For example, clusters that function at a very basic technological level will benefit from programmes aimed at basic vocational training and acquisition of more efficient machinery and equipment, while more advanced clusters will need support with education of engineers and financial support for R&D programmes.

4 An empirical illustration

Some of the principles discussed in the previous section will be illustrated with some case-study material. The focus is on farm equipment manufacturing, a highly important industry in developing countries when it comes to initial absorption and accumulation of knowledge about basic mechanical engineering principles. The industry essentially functions as a springboard from which capabilities subsequently diffuse to other industries as industrialisation gets under way (Johnston and Kilby, 1975).

The Pakistan Punjab is a fertile agricultural region known for its widespread adoption of modern cultivation practices since the late 1950s. Agricultural modernisation has been supported by a sizeable farm equipment industry which is organised into approximately eight major clusters, each consisting of 50–60 small-scale firms.[10] The industry emerged in the early 1960s in response to surging demand for irrigation equipment by farmers who were increasingly beginning to switch over to irrigation-based cultivation methods (ibid.; Aftab and Rahim, 1986). By the 1990s, the industry had evolved to the extent that it employed about 5,000 people and was capable of manufacturing well over 50 different items (Romijn, 1999). These included a number of relatively complex pieces of equipment with moving parts and high-grade steels, along with the simple rigid structures (such as mould board ploughs) that had been produced since the early stages of the industry's existence. Incremental design improvements had also been made to increase capacity, sturdiness, safety, and efficiency, although quality problems in production were still persistent (ibid.).

We first bring the CE and TC perspectives to bear to reflect on the evolution of the industry's performance and identify the best ways forward. Then we identify new policy insights by imposing our integrated approach on the case. The purpose of this exercise is not to come up with detailed policy prescriptions based on in-depth assessments according to the three approaches. Rather, the goal is to illustrate the broad differences in analytical perspectives taken and derive some important policy directions from them.

Starting with the CE perspective, the available evidence suggests that active networking and collaboration can be at best a partial explanation of the industry's growth. There is no tradition of active cross-firm horizontal collaboration in the production chain, as there is a fundamental lack of trust among competitors. Competition and associated secrecy are severe, even among people who belong to the same *bradri* (caste) (Nabi, 1988: 123).

Cooperation has worked somewhat better across different stages of the production chain, because these vertical relations are more complementary than competitive. For example, skilled machinists employed by parent firms reportedly pay frequent visits to subcontractors to ensure that new components are manufactured according to specification, and that materials are selected according to the parent firms' requirements (ibid., pp. 121–122). There is also some interaction with buyers of farm equipment. Some farmers provide performance feedback to the producers and suggest further improvements (Romijn, 1999). However, since the buyers are not themselves technologically knowledgeable, the impact of these linkages for firms' upgrading may not be all that dramatic. Also, the networking relations are not institutionalised in any formal way. There do not seem to be any bodies, such as industry associations or governmental agencies, that command sufficient trust and that are organisationally capable enough to be able to play a notable promotional role in this respect.[11]

If the past cooperation record is anything to go by, in the short run there is unlikely to be much scope for stimulating the industry's development through the kind of policy approach advocated in the CE literature. In the prevailing social-cultural environment it will doubtlessly prove hard to initiate and sustain active cooperative behaviour, especially among close competitors.

The TC approach draws attention to a rather different set of determinants to explain the industry's past growth. Firstly, strong demand-pull effects arising from agricultural modernisation provided incentives for innovation (Child and Kaneda, 1975). The industry emerged close to agricultural areas with fast-increasing crop yields. Rising purchasing power of farmers combined with increasing seasonal labour shortages fuelled massive investments in mechanisation, which in turn induced a critical minimum local demand for many new types of farm equipment. There was an obvious impetus for producers to introduce new product technologies by reverse-engineering prototypes imported by large engineering firms and wealthy farmers (Johnston and Kilby, 1975).

Supply-side factors were also important. The industry is endowed with a somewhat better institutional science and technology infrastructure than other sectors of small-scale industry. This has to do with its link with agriculture, a priority sector in the economy. There are national and provincial institutions employing engineers who are involved in adapting foreign farm equipment designs to local conditions. Their outreach to local firms has not always been effective, but some positive results have been noted (Romijn, 1999).

For example, several firms in Daska, one of the biggest clusters, were approached by extension officers from the national Farm Machinery Institute in Islamabad, who were looking for suitable partners with whom they could commercialise farm machinery prototypes developed by their institute. Interested workshops would receive in-house technical training and advice and would embark on a process of collaborative effort to iron out the teething problems of new equipment during field trials with local farmers (ibid., p. 243). Firms in another major cluster, Mian Channun, benefited from the establishment of a Dutch-funded training and common facility project on the local industrial estate, run under the aegis of the provincial Punjab Small Industries Corporation. Local firms received short training courses in heat treatment, the properties of different metals and their uses, and the use of jigs and fixtures. Moreover, they had access to specialised machining services (ibid., p. 243). Diffusion of skills was further assisted by technical and vocational training centres set up by the central and provincial governments (Aftab and Rahim, 1986).

From the perspective of the TC approach, policies for further upgrading would focus on the strengthening of these institutional features. For example, one could suggest improving the outreach and content of some of the existing technical extension and training programmes in order to make them more effective. Furthermore, the TC assessment would point to the importance of stimulating demand for the industry's products through measures supporting the continued advancement of the agricultural sector (see, for example, Romijn, 1999: 257–263).

Additional insights about the industry's development can be gained by applying the combined meso–micro framework outlined in table 6.1. One clear example relates to the favourable effect on firm-level technological effort emanating from transaction cost advantages and high spillovers in the local labour market. In particular, clustering of the farm equipment firms facilitated the establishment of specialised training suppliers who were attracted by a large local demand for their services. This did not just involve the government-run training services already mentioned above. Some reputable private firms in the industry also began to assume the status of training institutes. They even issued certificates as evidence that apprentices had completed their training there. The authenticity of these documents and the reputation of the firms in question could be verified easily in the local community, thereby reducing transaction costs for workshops looking to hire competent workers in the local labour market (mechanism IIb, A–D in table 6.1).[12] These specialised training suppliers also contributed to significant local labour-market spillovers (mechanism V-A) due to high inter-firm movement of trained labour (Stewart and Ghani, 1991: 585).

High spillovers in the product market constitute another example. Progressive farmers who import new farm equipment models are sources of information about new product designs. Manufacturers generally try to reverse-engineer foreign prototypes when they are passed on to them for repair and maintenance (Romijn, 1999: 196–197). Knowledge about new designs spreads rapidly due to co-location (mechanism V-A, C and D). Technology transfer appears to result predominantly from informal contact and observation, although marketing leaflets and the annual Horse and Cattle Show in Lahore (which features new locally produced farm equipment) may be of some importance as well (ibid., p. 226).

These observations illustrate some of the growth mechanisms that were identified in our framework analysing firm-level technological learning in clustered settings. Having these clearly in view enables us to derive more specific policy conclusions than would be possible on the basis of either one of the two partial approaches.

In particular, the examples suggest that, while the above-mentioned science and technology policies emanating from the TC approach already point in the right direction, these policies could be made more effective. This could be done, for example, by taking account of the local diffusion mechanisms and low transaction costs discussed above. Specifically, singling out progressive companies in the delivery of technical extension, consultancy services, and training programmes could be a more efficient way to deliver assistance than a broader-based effort aiming for direct coverage of large numbers of companies. This approach avoids the problem of having to mobilise large numbers of rather conservative entrepreneurs, often observed in more broad-based assistance programmes aimed at small-scale enterprises in poor countries. Many small entrepreneurs are difficult to convince about the benefits of participating in such formal assistance programmes, and are mistrustful of government involvement. Well-targeted interventions can help to create forerunners who could play the role of catalysts in spreading knowledge and progressive attitudes.

5 Conclusions

The analytical framework elaborated in this paper extends the collective efficiency approach that has been commonly used to analyse regional economic growth and SME competitiveness in development. This is done by opening the firm's black box, using well-known literature on firm-level technological capability acquisition in development. The resulting conceptual taxonomy sheds light on how firm-level capability building could be fostered through geographical clustering.

The main new policy insights that can be derived from our framework are summarised as follows.

First, the policies that are already part of the conventional collective efficiency approach, namely promotion of active networking and cooperation, are bound to become more effective when they are specifically aimed at stimulating technological learning, rather than covering all kinds of production-focused activities as well. Dynamic learning-focused collaborations are likely to have the most durable impact on industrial competitiveness. Thus, particular attention has to be paid to the positioning of local clustered firms in global value chains, as these chains hold potential for dynamic learning. The question as to which forms of chain governance can promote vertical diffusion of new technological and market knowledge through networks of local and non-local participants, thus stimulating firm-level capability-building processes, constitutes an important area of new research.

Second, the demand and supply policies to stimulate technological innovation that are typically part of the conventional technological capability approach become more effective when closely targeted at clusters rather than industries in general. The operation of the various agglomeration effects on firm-level technological effort and learning identified in this paper enhances the impact of these policies. Moreover, the mechanisms underlying technological learning in clusters could be taken into account in the design of the policies themselves. Notably, selective targeting of support to progressive firms is likely to pay dividends due to high spill-over effects.

In sum, the integrated approach suggests three different ways to provide more focus in policies to promote SME competitiveness. One lies in the object of the policies themselves (stimulating technological learning), the second is in the geographical coverage (concentrating on industrial clusters), and the third is in the implementation modality selected (targeting a few progressive firms).

This paper should be seen as a first step to design more effective policies promoting regional economic growth and SME competitiveness in development. The policy conclusions listed above clearly do not pretend to be more than broad guidelines for this purpose. Differences across sectors, levels of technological advancement, culture, social and political aspects, and so on, will give rise to heterogeneities that determine the concrete scope for feasible and effective policy interventions. Among other things, these contextual differences will affect the relative importance of the different mechanisms listed in table 6.1 and the specific content of these mechanisms. The taxonomy in the table can serve as a guideline for studying these differences. Further work should be aimed at operationalising the mechanisms in the taxonomy, so that it can be used as a tool kit for empirical research about firm-learning in clusters in practice.

Notes

1. Adapted from Bergman and Feser (1999).
2. See especially the case studies in the special issue of *World Development* on industrial clusters in developing countries (Vol. 27, No. 9, September 1999).
3. SMEs in developing countries also tend to lack a range of non-technical skills, for example marketing, financial, and personnel management. However, the technological capability literature has extensively pointed to the fact that weak technological and related organisational capabilities form particularly stringent constraints. Hence, we limit the focus of this paper to this aspect.
4. See especially the case studies in the special issue of *World Development* referred to in note 2.
5. There are many contributions to the capability literature. Lall (1992), UNCTAD (1996) and Herbert-Copley (1990) are good reviews.
6. The remainder of this section draws on Caniëls and Romijn (2003).
7. A fifth category is internal performance feedback, but Bell mentions that this activity by itself is unlikely to lead to substantial sustained learning.
8. In order to simplify the discussion we confine the focus to *direct* linkages, that is, effects which occur without intervening third variables which do not constitute technological efforts themselves. An example of an indirect linkage is cost savings in production which provide incentives for firms to expand, which in turn calls forth the need for new capabilities, and thus efforts to build them up.
9. The importance of this spillover has also been documented in many empirical studies, for example: Ahmad et al. (1984), Basant and Subrahmamian (1990), Cortes (1979), Fransman (1982), Johnston and Kilby (1975), and Nowshirwani (1977).
10. Small-scale in this context means that firms employ roughly between 5 and 50 workers.
11. Romijn, own fieldwork observations, 1994.
12. Romijn, own fieldwork observations, 1994.

REFERENCES

Aftab, K. and E. Rahim (1986) "The Emergence of a Small-Scale Engineering Sector: The Case of Tubewell Production in the Pakistan Punjab", *The Journal of Development Studies* 23(1): 60–76.

Ahmad Q. K., K. M. Rahman, K. M. N. Islam and M. E. Ali (1984) "Technology Adaptation and Employment in the Agricultural Tools and Equipment Industry of Bangladesh", World Employment Programme Working Paper No. 2-22/WP 134, Geneva: International Labour Organization.

Audretsch, D. B. and M. P. Feldman (1996) "Knowledge Spillovers and the Geography of Innovation and Production", *American Economic Review* 86(3): 630–640.

Baptista, R. (1998) "Clusters, Innovation and Growth: A Survey of the Literature", in G. M. P. Swann, M. Prevenzer and D. Stout, eds., *The Dynamics of Industrial Clusters: International Comparisons in Computing and Biotechnology*, Oxford: Oxford University Press, pp. 13–51.

Basant, R. and K. K. Subrahmamian (1990) *Agro-Mechanical Diffusion in a Backward Region*, London: Intermediate Technology Publications.

Bell, R. M. (1984) "'Learning' and the Accumulation of Industrial Technological Capability in Developing Countries", in M. Fransman and K. King, eds., *Technological Capability in the Third World*, London: Macmillan, pp. 187–209.

Bell, R. M. and M. Albu (1999) "Knowledge Systems and Technological Dynamism in Industrial Clusters in Developing Countries", *World Development* 27(9): 1715–1734.

Bergman, E. M. and E. J. Feser (1999) *Industrial and Regional Clusters: Concepts and Comparative Applications*, Web Book in Regional Science, Morgantown: Regional Research Institute, West Virginia University, online at: http://www.rri.wvu.edu/WebBook/Bergman-Feser/contents.htm.

Caniëls, M. C. J. (2000) *Knowledge Spillovers and Economic Growth*, Cheltenham: Edward Elgar.

Caniëls, M. C. J. and H. A. Romijn (2003) "Small-Industry Clusters, Accumulation of Technological Capabilities and Development: A Conceptual Framework", *Journal of Development Studies* 39(3): 129–154.

Ceglie, G. and M. Dini (1999) "SME Cluster and Network Development in Developing Countries: The Experience of UNIDO", paper presented at the International Conference on Building a Modern and Effective Development Service Industry for Small Enterprises, Committee of Donor Agencies for Small Enterprise Development, Rio de Janeiro, March 2–5, 1999.

Child, F. C. and H. Kaneda (1975) "Links to the Green Revolution: A Study of Small-Scale, Agriculturally Related Industry in the Pakistan Punjab", *Economic Development and Cultural Change* 23(2): 249–277.

Cortes, M. (1979) "Technical Development and Technology Exports to Other LDCs", in *Argentina – Structural Change in the Industrial Sector, Annex I*, Washington, D.C.: The World Bank, Development Economics Department.

Dahlman, C. J. and L. E. Westphal (1981) "The Meaning of Technological Mastery in Relation to Transfer of Technology", *The Annals of the American Academy of Political and Social Science* 458: 12–26.

DeBresson, C. and F. Amesse (1991) "Networks of Innovators: A Review and Introduction to the Issue", *Research Policy* 20: 363–379.

Dosi, G. (1988) "The Nature of the Innovative Process", in G. Dosi, C. Freeman, R. Nelson, G. Silverberg and L. Soete, eds., *Technical Change and Economic Theory*, London: Pinter, pp. 221–238.

Feldman, M. P. (1994) *The Geography of Innovation*, Boston: Kluwer.

Feldman, M. P. and R. Florida (1994) "The Geographic Sources of Innovation: Technological Infrastructure and Product Innovation in the United States", *Annals of the Association of American Geographers* 84(2): 210–229.

Fransman, M. (1982) "Learning and the Capital Goods Sector under Free Trade: The Case of Hong Kong", *World Development* 10(11): 991–1014.

Freeman, C. (1991) "Networks of Innovators: A Synthesis of Research Issues", *Research Policy* 20: 499–514.

Herbert-Copley, B. (1990) "Technical Change in Latin-American Manufacturing Firms: Review and Synthesis", *World Development* 18(11): 1457–1469.

Humphrey, J. and H. Schmitz (2000) "Governance and Upgrading: Linking Industrial Cluster and Global Value Chain Research", Working Paper 120, Brighton: Institute of Development Studies, University of Sussex.

――― (2001) "How Does Insertion in Global Value Chains Affect Upgrading in Industrial Clusters?", research paper, Brighton: Institute of Development Studies, University of Sussex.

Johnston, B. F. and P. Kilby (1975) *Agriculture and Structural Transformation. Economic Strategies in Late-Developing Countries*, New York: Oxford University Press.

Lall, S. (1992) "Technological Capabilities and Industrialisation", *World Development* 20(2): 165–186.

Malmberg, A. and P. Maskell (2002) "The Elusive Concept of Localization Economies: Towards a Knowledge-Based Theory of Spatial Clustering", *Environment and Planning A* 34: 429–494.

Marshall, A. (1920) *Principles of Economics*, London: Macmillan.

Mytelka, L. and F. Farinelli (2000) "Local Clusters, Innovation Systems and Sustained Competitiveness", UNU-INTECH Discussion Paper 2000-5, Maastricht: United Nations University, Institute for New Technologies.

Nabi, E. (1988) *Entrepreneurs and Markets in Early Industrialization. A Case Study from Pakistan*, San Francisco, Calif.: ICS Press.

Nadvi, K. (1999a) "Shifting Ties: Social Networks in the Surgical Instrument Cluster of Sialkot, Pakistan", *Development and Change* 30(1): 141–175.

――― (1999b) "The Cutting Edge: Collective Efficiency and International Competitiveness in Pakistan", *Oxford Development Studies* 27(1): 81–107.

Nadvi, K. and G. Halder (2002) "Local Clusters in Global Value Chains: Exploring Dynamic Linkages Between Germany and Pakistan", IDS Working Paper 152, Brighton: Institute of Development Studies, University of Sussex.

Nelson, R. R., ed. (1993) *National Innovation Systems. A Comparative Analysis*, Oxford: Oxford University Press.

Nelson, R. R. and S. G. Winter (1982) *An Evolutionary Theory of Economic Change*, Cambridge, Mass.: The Belknap Press of Harvard University Press.

Nowshirwani, V. F. (1977) "Employment, Technology Transfer and Adaptation: The Case of the Agricultural Machinery Industry in Iran", World Employment Programme Working Paper No. 2-22/WP 31, Geneva: International Labour Organization.

Richardson, H. W. (1978) *Regional and Urban Economics*, Hindsdale: Dryden Press.

Romijn, H. (1999) *Acquisition of Technological Capability in Small Firms in Developing Countries*, London: Macmillan.

――― (2001) "Technology Support for Small-scale Industry in Developing Countries: A Review of Concepts and Project Practices", *Oxford Development Studies* 29(1): 57–76.

Schmitz, H. (1995) "Collective Efficiency: Growth Path for Small-scale Industry", *Journal of Development Studies* 31(4): 529–566.

――― (1999) "Global Competition and Local Cooperation: Success and Failure in the Sinos Valley, Brazil", *World Development*, 27(9): 1627–1650.

Schmitz, H. and P. Knorringa (2000) "Learning from Global Buyers", *Journal of Development Studies* 37(2): 177–205.

Schmitz, H. and K. Nadvi (1999) "Clustering and Industrialization: Introduction", *World Development* 27(9): 1503–1514.

Stewart, F. and E. Ghani (1991) "How Significant Are Externalities for Development?", *World Development* 19(6): 569–594.

Swann, P. (1998) "Towards a Model of Clustering in High-technology Industries", in G. M. P. Swann, M. Prevenzer and D. Stout, eds., *The Dynamics of Industrial Clusters: International Comparisons in Computing and Biotechnology*, Oxford: Oxford University Press, pp. 52–76.

Tewari, M. (1999) "Successful Adjustment in Indian Industry: The Case of Ludhiana's Woollen Knitwear Cluster", *World Development* 27(9): 1651–1671.

UNCTAD (1996) *Fostering Technological Dynamism: Evolution of Thought on Technological Development Processes and Competitiveness. A Review of the Literature*, Geneva and New York: United Nations.

――― (1998) "Promoting and Sustaining SME Clusters and Networks for Development", paper presented at the Expert Meeting on Clustering and Networking for SME Development, UNCTAD Secretariat, Geneva, September 2–4, 1998.

Visser, E.-J. (1999) "A Comparison of Clustered and Dispersed Firms in the Small-Scale Clothing Industry of Lima", *World Development* 27(9): 1553–1570.

Von Hippel, E. (1988) *The Sources of Innovation*, New York: Oxford University Press.

7
The contribution of skilled workers in the diffusion of knowledge in the Philippines

Niels Beerepoot

1 Introduction

Increasing attention has been given recently to collective learning or localised learning as a source of competitiveness for regional concentrations of (small) firms. The creation and further development of a base of shared knowledge among individuals making up a local productive system are recognised as a prerequisite for competitiveness, precisely because of the drive towards globalisation and the homogenisation of formerly diverse factors of production. Based on fieldwork conducted among small enterprises in the Philippines, this paper concentrates on two issues in the discussion on learning in small-enterprise clusters. The first issue is how horizontal and vertical inter-firm relations serve as preconditions for knowledge accumulation in industrial clusters. How these relations enable or hinder the development of a local knowledge base will be identified. The second issue is the role of skilled workers in the creation and diffusion of knowledge in small-enterprise clusters. When created knowledge is embodied in a regional industrial structure – in the form of procedures, routines, and the building of a common production culture – it is often difficult to identify exactly where in the labour force the knowledge and skills are embodied and how embodiment has taken place. The labour force is often treated in an undifferentiated fashion despite the recognition skilled workers get as carriers of the region-specific knowledge by which communities gain their competitive advantage. How skilled workers play a role in the local diffusion of knowledge, and which forms of knowledge are transmitted more easily than others, will be identified in this paper. To this end, empirical fieldwork was carried out in a cluster of furniture producers in Cebu (the Philippines) between November 2001 and March 2002.[1] By means of both qualitative and quantitative research techniques, information was gathered among managers *and* workers in this cluster to identify what contribution a poorly educated but highly skilled workforce has in the development of the industry.[2]

Section 2 gives a brief introduction to the current debate on learning processes as a key source of competitive advantage for industrial clusters. Attention is paid to the learning indicators in developing country clusters. Many such clusters operate in traditional, low technological, industries with low levels of education among members of the workforce, which negatively affects localised learning processes. While there is already a growing body of literature on localised learning processes in clusters in industrialised countries, evidence about learning processes in a developing country context is still poor. Section 3 explores the role and importance of skilled workers in localised learning processes. Section 4 gives an introduction to the furniture cluster in Cebu. Section 5 examines how practices and codes of conduct in this cluster stimulate learning processes. Section 6 then examines the role skilled workers play in the creation and diffusion of knowledge within this cluster. Finally, section 7 offers some concluding remarks on the paper's findings.

2 Localised learning in small-enterprise clusters

2.1 *Conceptualising learning*

Learning involves the development of technological, social, organisational, and financial knowledge embedded in the labour force (Capello, 1999; Maskell et al., 1998). The rate of learning determines the speed of economic change and the kind of learning determines the direction of economic change. Those firms, sectors, regions, and nations that can learn faster or better (higher quality or cheaper for a given quality) become competitive because their knowledge is scarce and therefore cannot be immediately imitated by new entrants or transferred, via codified and formal channels, to competitor firms, regions, or nations (Storper, 1997). Sustained competitiveness therefore results from the ability to create, obtain, and utilise knowledge more quickly than competitors (Lawson, 2000). This is one of the main reasons for the recent interest in learning and innovation in a period where competitive advantage is to a large extent derived from investments in human resources. Learning as a tool for regional competitiveness opens new opportunities for development strategies in underprivileged countries.

For small producers, clustering and cooperation can be the major factors facilitating a degree of efficiency and flexibility in production that (small) individual producers can hardly attain. The presence of similar producers in the same region can stimulate and facilitate learning processes. A general finding from recent studies (see, for example, Audretsch and Feldman, 1996; Caniëls, 1999) is that new knowledge can be transmitted easiest over short distances. Spatial proximity between actors in various ways makes easier those knowledge spillovers and interactions which form the basis for innovation and learning, and it is in this context that spatial clustering becomes a key focus of analysis (Malmberg and Maskell, 2002). Collective initiatives by small producers utilising available information and communication can stimulate the generation of a local knowledge base. Low transaction costs in clusters directly facilitate (horizontal and vertical) business interaction, joint projects, and labour mobility, which are the main vehicles through which skills, knowledge, and ideas travel across firms (Caniëls and Romijn, 2001). The interaction between economic, socio-cultural, political, and cultural actors in a given place may trigger learning dynamics and enhance the ability of actors to modify their behaviour and find new solutions as their competitive environment changes (Malmberg and Sölvell, 1997).

For Lawson (2000) learning is the process by which the success and failure that individuals experience become encoded into routines and practices of the community of which they are a part. A deeper understanding of the nature of learning processes is needed to understand the contribution learning can make to cluster development. Learning, whether by individuals, firms, regions, industries, or nations, is largely treated in an undifferentiated fashion in the economic literature (Lawson, 2000). Who are subject to changes in their knowledge and capacities and the area in which a local knowledge base is built up remain unclear in many studies. It is evident that it is individuals who learn rather than some organisation or group. Conceição and Heitor (1999) make a distinction between different kinds of learning that can lead to the accumulation of knowledge. Learning as a formal process is through education and training (knowledge in the forms of skills) and R&D efforts (accumulation of ideas). Learning as an informal process is through experience (learning by doing) and (social) interaction. In the latter process, the emphasis is given to the formation of personal and professional contacts that result from social interaction. The study of informal learning processes is more complex and less amenable to empirical testing (Conceição and Heitor, 1999). According to Lawson (2000), shared languages, knowledge, conceptions, and identities are crucial conditions for learning and knowledge acquisition to take place, underlining that learning is both active as well as context dependent. Knowledge transmission can take place through various mechanisms. Skilled labour mobility within the local labour market, various kinds of customer–supplier relations as well as relations to all kinds of service firms, and new start-ups through direct or indirect spin-off activity feature as crucial channels whereby learning takes place. Perez-Aleman (2000) also emphasises that learning is a context dependent process. In her opinion, learning is an interactive and socially embedded process that depends on an institutional context that builds reciprocity and collaboration. A local culture with specific norms, values, and institutions (formal and informal) makes it possible to transfer tacit forms of knowledge from one actor to another (Malmberg and Maskell, 2002).

2.2 Context for learning in developing country clusters

Various studies (for example in *World Development* special issue, Vol. 27, No. 9, September 1999) have already paid attention to small-enterprise clusters in a developing country context. Some clusters in developing countries display a growth potential that goes beyond informal survival strategies and indicates localised competitiveness based on increasing specialisation among small firms (Humphrey and Schmitz, 1996). While there are already a number of studies on clusters in developing countries, there are still very few studies on learning and innovation in lower- and middle-income countries. If and how local circumstances enable the development and diffusion of knowledge in such a different social and institutional context and who the beneficiaries are have not as yet been given much attention.

Many clusters in developing countries are primarily artisan-based or operate just beyond artisan-based production with poorly educated production workers who mostly use simple equipment. A limited willingness or possibility to invest in their human resources constrains the capacity of such small enterprises to absorb and accumulate knowledge. The abundance of cheap labour in many developing countries provides a disincentive to invest in mechanisation and skills development. Investing in workers' skills and capacities tends to become a zero-sum game in clusters because entrepreneurs never know if other entrepreneurs are willing to equally invest in training. The uncertainty about returns on investments in skills upgrading, the fear of piracy, and the inter-firm movement of skilled workers are often cited as reasons why many employers invest so little in training activity.

Most industrial clusters in developing countries also operate in so-called "buyer-driven" value chains. The notion of "buyer-driven" value chains expresses the idea that the buyer exercises control over the chain even in the absence of ownership (Gereffi, 1999; Humphrey and Schmitz, 2000). Opportunities for upgrading are basically driven by the requirements of the buyer and their willingness to transmit knowledge to their suppliers. Global value chains can therefore create barriers to local upgrading because global buyers protect their key strategic functions (for example design, branding, and marketing) in the chain. According to Humphrey and Schmitz (2000), there are three reasons why developing country clusters are less likely to operate in chains characterised by innovation-conducive network-based relationships. First, the scope for learning is much greater at early stages in the product life cycle: most industrial exports of developing countries are typically mature products. The second reason is the competence differential between buyers and producers: network-based relations with (foreign) clients are rare according to Humphrey and Schmitz. The third reason is that information-rich collaboration between firms is often temporary in nature and relations soon tend to become market-based.

Some authors (for example MacKinnon, Cumbers and Chapman, 2002) claim that lack of economic development reflects the poor learning characteristics of a region. A lack of trust among various local actors can prevent the development of a common language and culture that are needed to facilitate the exchange of knowledge in a region. The idea of establishing dependency ties with other enterprises is not considered as attractive by firms as it reduces control and authority (Pietrobelli and Barrera, 2002). Many clusters in developing countries also have a strong internal hierarchy whereby a few leading families own the largest, more modern factories and dominate the industry through the local business associations. These actors try to monopolise benefits and become a source of conservatism instead of innovation even though they have the financial capacity to invest in upgrading (Knorringa, 2001).

The study of learning processes in a developing country context should focus more attention on how local circumstances or milieux enable or hinder the development and diffusion of knowledge. Knowing why some industrial clusters in developing countries are "learning-rich" while others are "learning-poor" would make an important contribution to understanding their long-term dynamism (Bell and Albu, 1999). Endogenous regional development as emphasised by Pyke and Sengenberger (1992) concerns a greater self-reliance on economic development through the mobilisation of existing resources and self-created local organisations. The local governance structure that is often given recognition in localised learning processes is far less developed in many developing countries. Many clusters in developing countries operate not only in less innovative, mature industries, but also often in industries with predominantly artisan-based production. This industrial structure, with often only limited local differentiation and specialisation among producers, does not provide the necessary environment for localised learning. In industries with very simple technology and an abundant supply of cheap labour, not only can workers be trained at minimal cost on the job, but there are also no incentives to invest in more skills-intensive technologies (Bennel, 1999).

3 Skilled workers, embodied expertise, and localised learning

While there has been a growing body of literature on the existence of collective learning processes in small-enterprise clusters, studies on the role of skilled workers in the local development and diffusion of knowledge are still limited. A learning region requires a human infrastructure of workers who can apply their intelligence in production (Florida, 1995). Knowledgeable entrepreneurs and adaptable well-trained labour are the key to both the promotion of flexible, innovative, quality conscious, and productive enterprises and the future employability and security of workers (ILO, 1999). So far, little attention has been paid to the position of workers in clusters and the contribution of skilled workers in a localised learning process. Sandee (1995) suggests that there is a need to realise that clusters do not constitute a homogeneous workforce and the impact of innovation adoption may lead to increased differentiation between groups of workers.

In an increasing number of studies, mobile workers are recognised as carriers of knowledge in the local labour market. The knowledge and know-how of workers are the intangible assets that can make regions competitive in international markets. The distinctive attribute of industrialised country "high-learning" regional clusters is directly associated with these regions' ability to retain and expand the regional pool of skilled workers and entrepreneurs on which development of collective learning capacity crucially depends (Keeble, 2000). In industrial clusters in developing countries, formal training suppliers have a limited role in skills development. Formal or institutionalised training contributes only in a very small way to skills development and employment in developing countries (Singh, 2000). Non-formal or informal mechanisms are more important in the local diffusion of knowledge in clusters with predominantly artisan-based production. Expertise might be handed down "through the community", from father to son, mother to daughter, and from colleague to colleague, so that it forms part of a long-standing cultural heritage for the region (Pyke and Sengenberger, 1992). The tacit knowledge concept is useful in describing the difficult-to-transfer local competencies by which communities gain their competitive advantage. With tacit knowledge one learns procedurally by imitating the observed behaviour of one or more masters in a "community of practice" and trying it out in practice subject to correction by that community (Nooteboom, 1999).

According to Storper (1997), learning in low technology industries can generate more high-wage jobs. This corresponds very well to the distinction Pyke and Sengenberger (1992) made between the "low road" and the "high road" in international competition. The first refers to seeking competitiveness through low labour costs, cheap materials, and a deregulated labour environment. The alternative is competition based on efficiency enhancement, innovation, high quality, functional flexibility, and good working conditions. More specifically focused on upgrading in artisan-based industries, Knorringa (1995) uses the term "premium artisanal employment" to describe a situation in which workers are offered greater regularity and enjoy better conditions as quality conscious entrepreneurs aim to utilise the surplus value of cheaply available artisan skills. However, Knorringa emphasises that more successful industrial clusters do not necessarily follow higher-road employment strategies, nor do more successful entrepreneurs pay higher wages. Evidence from Indonesia (Sandee, 1995) indicates that buyer-driven innovation processes in Indonesian clusters have stimulated piece-rate payment systems and the development of out-contracting among producers.

4 The furniture industry in Cebu

The furniture industry is a mature industry that is present in virtually all developing countries. From an international perspective, furniture making is still a very fragmented industry. The majority of production takes place in small enterprises using only local resources. The entry-level in this industry is low, particularly in developing countries. The furniture industry in the Philippines has its origin in local handicraft production. For centuries artisans used simple tools and locally available materials to produce consumer goods for the local market. Craft-based production is therefore deeply embedded in the Philippine society. Skills in the craft sector are basically learned on the job through an older worker teaching the novice over a relatively long period of time (Fluitman, 1989). This transmission of crafts requires a high level of trust and a common work ethic and production attitude. Kinship and a common culture are important in the diffusion of skills and knowledge among workers.

Although craft-based products had already been produced in cottage industries for the export market during the period of colonial rule, the export demand for these products increased greatly in the 1950s. There were several reasons why Cebu developed as the prime location of furniture making in the Philippines. The development of the industry in Cebu benefited from its close proximity to rattan, which has long been the main raw material. In combination with good port facilities and a relatively well-developed entrepreneurial middle class, Cebu was able to develop as the prime producer of furniture for export, accounting for approximately 60 per cent of total Philippine furniture exports. Currently there are around 175 exporters of furniture in Cebu. A rough estimate by close observers is that around 45,000 people find direct or indirect employment in the industry. The furniture industry contributes approximately 3 per cent to Philippine GDP. It is the largest foreign exchange earner among the more "traditional" industries in the Philippines, although their exports are quite small when compared with those of such high technology categories as machinery and electronic manufacturing (Berry and Rodriguez, 2001). However, the Philippine furniture industry is losing market share to its Asian neighbours. In the 1970s and early 1980s the Philippines was the prime producer of furniture in Southeast Asia. Nowadays this position has been taken over by China and Indonesia: in some cluster studies (for example Pietrobelli and Barrera, 2002; Rabellotti, 1999) it is briefly mentioned that it is likely that the supply of cheap products from China will erode the position on the world market of many developing country clusters producing shoes, garments, or furniture. The current share of Philippine furniture makers in the world market is less than 1 per cent. A diminishing supply of raw materials and relatively high labour costs compared to neighbouring countries are often cited as the main reasons for losing competition. Under such conditions it is necessary to identify to what extent an industrial strategy can be based on localised learning. The effective utilisation of the industry's greatest asset, a highly skilled flexible workforce, is then the greatest challenge in this development trajectory.

5 Enabling environment for learning in Cebu's furniture industry

The most important aspects of the localised learning capacity are common regional culturally based rules of behaviour, a language of engagement and collaboration, and tacit codes of conduct between firms, which enable the development of trust, itself essential for innovative collaboration (Keeble, 2000). Learning is a socially embedded process that cannot be understood without taking into consideration its institutional and cultural context (Lundvall, 1992, cited in Asheim and Cooke, 1999). As

Schmitz (1999) highlights, trust is essential for a deep division of labour and cooperation between firms to be effective at reasonable cost. Based on open interviews with entrepreneurs and local support organisations, this section describes how inter-personal and inter-firm relations enable the development and diffusion of knowledge in the furniture cluster in Cebu.[3] How milieux in developing country clusters as mentioned in section 2.2 facilitate or constrain knowledge accumulation and learning processes in Cebu will be identified.

Like most industrial clusters in developing countries, the furniture cluster in Cebu is a low technological industry. Many furniture exporters in the Philippines have a strong foothold in the informal sector through the outsourcing of work to small subcontractors and home-workers. Multi-level subcontracting is practised as a way of "taking the heat off" a particular corporation. For example, a Manila-based company subcontracts to regional agents who, in turn, often mobilise women and children on low pay and under poor labour conditions (McFarlane, 1998). This practice, with a predominance of piece-rate payment of workers, is generally accepted among producers in the Philippines. As one handicraft company promotes itself on its website: "from a company of 5 people we have come a long way. We have 60 full time employees and over 120 in-house artisans (contractual labourers). We also have a production support base of over 2500 families". This simple example gives an idea of the division of labour in the industry and the subsequent job quality and security of tenure of workers in the different stages of production. At the same time it becomes clear that workers in the lower strata of the production hierarchy do not have much access to formal upgrading of skills and the development of tasks. It should be emphasised that many tasks in the furniture industry in Cebu (i.e. manual sanding and polishing) require only little training and can be done after only a short introduction on the job. The opportunity to accumulate knowledge is limited in such a low technological position.

Export companies in Cebu concentrate themselves on the final stages of production (i.e. finishing and packaging), while an increasing number of tasks are sourced out to subcontractors. Flexibility and control of the company are the main reasons for this practice. These companies, which work for international buyers (who supply them with the designs and product specifications), at least try to protect their own stage in the value chain as well as possible. When, in the direct aftermath of the September 11 events in 2001, the Philippine furniture industry was confronted with a sudden collapse in orders from the United States (their prime export market), the local outsourcing strategy certainly helped exporters to survive. Most of the hardship was felt at the lower levels of the local production hierarchy. Companies claim that when they are too big they become vulnerable to threats of unionisation. Subcontracting is therefore primarily a cost-cutting and risk-spreading measure rather than a way to improve product quality through specialisation of tasks. Under such conditions cooperation along the value chain to stimulate the exchange of ideas and development of new knowledge is difficult to identify. Relationships between furniture exporters and their subcontractors are based on dominance by the former rather than equality or seeking active involvement of subcontractors in discussion on technical issues or the development of products.

Although a number of the exporters in Cebu once started as subcontractors themselves, many of them have a low view of their subcontractors. Complaints about late deliveries and inconsistent quality are often heard from exporters. They seem to accept these struggles with their subcontractors, however, instead of providing them with extra training or otherwise producing the goods in-house themselves. An expression of how exporters try to establish their dominance over the subcontractors is through payment with post-dated cheques, a system under which the subcontractor receives his payment only 30 or even up to 60 days after the delivery of the goods. The International Labour Organization (ILO, 1999) strongly associates the growth in employment in small enterprises in developing countries with a parallel trend of a growth in non-standard forms of informal and flexible employment relationships. These reduce the access of workers to a range of economic and social entitlements as well as increasing insecurity. Bennel (1999) calls this process the "informalisation of the formal sector" in order to exploit cheap labour. This trend is in sharp contrast to practices found in some successful clusters in Europe. According to Keeble (2000), big firms in various successful European districts have, through positive subcontracting relations with local SMEs, been influential in creating a regional culture of trust and collaboration which encourages innovation.

While producers try to obtain dominance over their subcontractors along vertical relations, the establishment of horizontal relations is similarly difficult. Cooperation between furniture companies that goes further than simple and incidental collaboration is rarely seen in Cebu. The significance of a common culture is often stressed in the development of clusters. However, despite a common culture, diversity in the social backgrounds of entrepreneurs can hinder collaborative efforts. Also a lack of homogeneity among firms in terms of capacity, equipment, work processes, and quality level hinders joint action. There are small furniture exporters in Cebu who ship only two containers of furniture a month, while a few leading exporters ship up to 50 containers every month. The potential threat of surrendering independence is given more weight by many entrepreneurs in Cebu than the potential benefits of inter-firm specialisation. The evidence from Cebu corresponds very well to research being done elsewhere in developing countries (see, for example, Rabellotti, 1999) where there are also few strong horizontal ties. While (selective) cooperation in many clusters takes place along vertical axes, competition takes place on the horizontal level. The market failure in learning, i.e. the uncertainty about returns on investments in skills upgrading and networks of association, makes horizontal inter-firm cooperation difficult to establish. The fear of piracy and the inter-firm movement of skilled workers are important reasons why horizontal cooperation is limited in Cebu.

The evidence from Cebu indicates that conscious "positive" mechanisms of knowledge transfer, i.e. through vertical and horizontal inter-firm collaboration and specialisation and investment in training activity, rarely play a role in the diffusion of knowledge in the Cebu cluster. Alternative mechanisms may have a greater role in a localised learning process. Operating in a cluster provides companies with the opportunity to monitor the work of similar firms and combine this with their own efforts. Entrepreneurs and designers indicated in interviews that they often look at new products of related sectors, such as home accessories, to pick up ideas for new materials that can be used for furniture as well. The piracy of skilled workers and stealing of ideas or product designs can be labelled as conscious "negative" mechanisms of knowledge transfer. Some entrepreneurs in Cebu had no difficulty in saying that they are usually not the first locally to come up with new products or designs, but look to see which products by other local furniture makers sell well and then produce a similar item. Information about this is usually acquired through local furniture fairs, by shopping around in subcontracting units, or from workers who transfer between companies. These processes play an important role as alternative mechanisms of knowledge diffusion. Despite their important role as alternative learning mechanisms, these processes are to a great extent responsible for the low level of trust that exists among producers in this cluster.

6 Skilled workers and learning in Cebu's furniture industry

A main problem with studying learning processes is to identify exactly where in the local labour force knowledge and skills are embodied and how this embodiment has taken place. It is necessary to identify which groups in the labour force play a role in the local diffusion of knowledge. The collective aspect of learning sometimes comes up somewhat naively in enthusiastic usage of the "learning region" metaphor, as if everyone involved in them aims at the joint goal of creating competitive advantage, as if learning regions were communities happily learning collectively with no signs of friction or domination (Oinas and Malecki, 1999). In an insecure production environment, prevalent in most developing countries, it can be expected that skilled workers are concerned with the protection of their scarce knowledge. The transmission of knowledge to others can weaken their own position in the local labour market.

The difficult relations between exporters and subcontractors have already been mentioned in the previous section. In many instances exporters weaken their subcontractors by pirating their most skilled workers. These workers are asked to work directly for the exporters at higher wage levels. In recruiting workers, companies give more importance to experience and dexterity than to diplomas. A skills test in which the applicant demonstrates his talent is a central part in the application procedure. Working conditions in the furniture industry are tough as most employees have to work 6 or even 7 days a week for at least 8–10 hours per day. The furniture industry is the main employer in Cebu for people with only limited education and few employment opportunities. Most of the production workers only have elementary or (incomplete) high school education. Employment in an informal production environment, where skills are learnt on the job, is often the only opportunity for them. Despite their limited education, these people are often highly skilled in their specific craft and display a sound knowledge of the materials they work with. Utilising the full potential of their knowledge and skills is a great challenge in a regional learning process in artisan-based clusters in a developing country context.

To determine whether local diffusion of knowledge takes places and who plays a role in this process, it is necessary to carefully classify the local labour market. A survey was held among 280 workers from 50 companies in the export furniture industry in Cebu. These 50 companies were randomly selected from the databank of the Cebu Furniture Industry Foundation. Only rough estimates exist of the total number of workers in the industry (there are around 45,000 workers). As in many artisan-based industries with a strong foothold in the informal sector, workers enter and leave the industry easily. The survey tried to capture the diversity of workers in the local furniture industry. Workers were asked questions about their current and previous position(s), occupational status, educational background, additional training, number of years in the industry, local labour mobility, and who taught them their skill. Additional information for classification was collected through open interviews with entrepreneurs in the furniture cluster and informal, open interviews with subcontractors. Interviews with entrepreneurs concentrated on topics such as intra- and inter-firm relations, sources of knowledge, and investments in training and skills development. Based on data collected from various sources, the labour market categories of workers are given in table 7.1.

The table demonstrates that the local labour market is more segmented than is often identified in clustering studies. The industrial cluster with a high level of artisan-based production, at the same time operating in international value chains, employs both people who learned their skill primarily on the job as well as people with formal education related to their job. Evidence from the fieldwork in Cebu indicates that a distinction should be made between "knowledge protectors" and "knowledge transmitters". Knowledge protectors are people who have a relatively secure position in the local labour market and who are very much concerned with defending this secure position against others. This group mainly includes entrepreneurs, production managers, and designers. The protection can be against subcontractors, fellow designers, production managers, and other entrepreneurs. A characteristic of this protectionist behaviour is a limited willingness to share knowledge with other people within the same category or to initiate collective activity. These people protect their specific knowledge as a scarce good on the local labour market. The scarce knowledge or skills of these privileged persons can be on the production side (with specialised production and design skills) or on the operational side (for example external contacts, knowledge of production organisation, or contacts with subcontractors). Joint efforts that should stimulate localised learning processes and help in the diffusion of knowledge are limited among these groups.

Entrepreneurs and production supervisors have a limited willingness to invest in workers' skills, training, and technological capacities. The fear that production workers will be pirated, or set up companies themselves, often prevents companies from investing in training. This limited attention to training and skills development is a general trend in developing country clusters, where price-based competition is still predominant. Several entrepreneurs in Cebu asked lead men in their companies (mostly people with a position in category three – ("Skilled, secure/insecure workers" – in table 7.1) to start companies themselves and work exclusively as a subcontractor for their current employer. The advantage for the entrepreneur is increased flexibility in production and lower costs because more production now takes place in an informal production environment. The second advantage is that the loyalty of the subcontractor, based to a large extent on dependency, prevents the leaking of knowledge or stealing of ideas. Through this process the company can outsource production activity but still retain control of key knowledge. The subcontractor is in this context so much weaker and dependent that it can

Table 7.1 Segmentation of the labour market in the furniture industry in Cebu

Category	Special skilled, secure entrepreneurs and workers	Skilled, secure workers	Skilled, secure/insecure workers	Semi-skilled, insecure workers
Position	Entrepreneurs, production managers, designers	Supervisors, lead men in big companies	Lead men, sample-makers, skilled workers	(Semi-)skilled, work for subcontractors, apprentices
Education	College degree	College degree or vocational course	High school graduate, (some) college	Elementary graduate, (some) high school
Occupational status	Regular employee	Regular employee	Regular + contractual	On-call, job-outer, piece-rate workers
Source of knowledge	Formal training, then experience	Experience then formal training	On the job, experience	On the job
Additional training	Yes (some)	Yes (some)	No	No
Job security	Medium	Medium	Medium/low	Low
Payment[a]	>150% of minimum wage	Up to 150% of minimum wage	Minimum wage	Below minimum wage
Local labour mobility	Medium	Medium	High	Very high
Knowledge transmitter	No	Little	Yes	No (receiver)
Knowledge protector	Yes	Yes	No	No

Source: Own survey results and open interviews, 2002.
Note: The last two points in the table are based on a combination of survey answers (on sources of knowledge) and information from the open interviews with entrepreneurs and subcontractors. Because the table is based on both quantitative and qualitative data, and in some cases the borders between the categories are unclear, it is difficult to give a number of surveyed workers per category.
a. The minimum wage in Cebu at the time of the fieldwork was 195 Philippine peso (= US$3.75) for an 8-hour working day.

be questioned to what extent he can serve as a source of knowledge or feedback for exporters.

Knowledge transmitters on the other hand are generally lower in the local production hierarchy. This group includes people who have limited formal education and learned their skills primarily on the job. Despite their limited education, these production workers are often highly skilled in weaving or wood carving. Skilled workers often have an apprentice or assistant under their guidance who helps them with production. Piece-rate payment of skilled workers generally encourages workers to have a (young) helper with them. This practice is an example of the transfer of skills through informal mechanisms. Entrepreneurs admit that they have little control over this process of transfer of skills among workers. The on-the-job learning period may take just one month for simple rattan weavings but up to three years for mastering the skill in sophisticated wood carving. The transfer of skills among workers is limited to that directly needed in the production unit. The craft that a young apprentice takes up depends on the area he lives in and the skills that his relatives have. Workers also very much stick to the skill that they know and rarely consider whether acquiring more skills would improve their chances of future employability. There are very few workers in this cluster with multiple skills: most workers just have expertise in one specific craft or with one material.

The availability of a large surplus of skilled production workers in the local labour market is generally encouraged through their easy willingness to share knowledge and teach others their skills. Most of the surveyed production workers indicated that they learned their skills primarily from other workers, their lead men, or relatives. For lower hierarchy workers, apprentices or helpers are not seen as a threat to their own position in the labour market. The prospect that eventually these helpers will look for a skilled position has not made them more hesitant to share their knowledge. The high labour mobility of these workers within the local industry can bring benefits to the entire industry as knowledge becomes more widely accessible. Skilled workers may not always have a secure position within the company they work for, but they feel that their knowledge and skills give them the opportunity to easily find employment elsewhere in the local industry. Workers move especially easily between subcontracting units. In some cases they are hired only when there is an order and move afterwards to another subcontracting unit in the same neighbourhood for a similar job. It can be questioned if a high labour turnover can distort the build-up of distinctive skills in individual companies within the Cebu cluster. Undifferentiated, basic production knowledge and techniques are widely available within this cluster, but complementarity of knowledge and skills necessary to strengthen the industrial base is limited.

When workers and subcontractors work under circumstances with limited security of tenure, like most workers in the "Skilled, secure/insecure workers" and "Semi-skilled, insecure workers" categories in table 7.1, they seem to be more willing to share knowledge and undertake common endeavours. For the interviewed subcontractors, discussions with other subcontractors were one of their key sources of market information. These contacts are important because formal providers of information or knowledge from outside the cluster are difficult to access. For subcontractors the necessity to protect their specific knowledge does not have much importance. Localised learning and interventions to stimulate learning might therefore have most success at lower levels of the production hierarchy. On this level, production knowledge and skills are already transmitted easily. When companies expand or people enjoy relative security in the local labour market, the willingness to associate or undertake collective activity that should encourage knowledge accumulation diminishes.

7 Conclusion

The aim of this paper was to contribute to a better understanding of learning processes in small enterprises by concentrating on the preconditions for a local enabling environment and the role of skilled workers in localised learning and diffusion of knowledge. In existing literature it is stated that the scope for learning is much smaller in developing country clusters because they mostly operate in mature markets dominated by global buyers. Upgrading then depends to a large extent on the buyer's willingness to transmit knowledge.

The evidence from fieldwork in Cebu indicates that localised learning in the furniture cluster is negatively affected by local modes of production organisation. Local value chain relations and outsourcing strategies are based on dominance by exporters instead of seeking active involvement of subcontractors. Protection of their own position in the value chain hinders the development of both horizontal and vertical ties among large producers in Cebu. Learning and the development of a local knowledge base in this cluster are also negatively affected by the increasing outsourcing of work to the informal sector and through piece-rate payment schemes for workers. An increasing informalisation of production activity hinders the development of a regional culture of trust and collaboration, which is a necessary pre-condition for localised learning processes. The informalisation of production also negatively affects the investment in skills and capacities that is necessary for upgrading the furniture industry in Cebu.

The furniture industry in Cebu provides employment to many people with sophisticated skills but limited formal education. The majority of the workers in the furniture industry have learned their skills through informal mechanisms, with only a small number of workers depending on formal training for their knowledge and skills. In this cluster, learning processes and knowledge accumulation are negatively affected when formally trained, relatively secure, workers and entrepreneurs are very much concerned with protecting their scarce knowledge against fellow entrepreneurs, designers, production managers, and subcontractors. This protectionist behaviour among these key personnel hinders a regional learning process and the accumulation of knowledge. At the same time production skills and knowledge are transferred more easily among insecure production workers as part of a longstanding cultural heritage in the region. The negative consequence of this strong reliance on informal learning mechanisms is that distinctive and complementary skills that strengthen the entire cluster are hardly built up.

In an attempt to better understand localised learning processes, it is necessary to identify how a learning process is built up and who is involved in it. Future research can concentrate on further disentangling localised learning processes. The ease with which knowledge is transmitted does not necessarily depend on the tacitness of the knowledge, but more on the position of its owner and how this person values his or her knowledge. The second issue on which future research could concentrate is who plays a role in the local diffusion of knowledge. A better understanding of which people in the labour force are involved is necessary for developing interventions to stimulate localised learning. The success rate of interventions will largely depend on the willingness of individuals to share knowledge and their capacity to build further on this.

Notes

An earlier version of this paper was published in the *International Journal of Technology and Globalisation*.
1. This paper is part of the research for a Ph.D. thesis which concentrates on the accumulation and diffusion of knowledge in low-technological clusters in the Philippines and the contribution of skilled workers to these processes.
2. A survey was held among 280 production workers representing 50 companies. Additional information was gathered through open interviews with entrepreneurs in this cluster.

Open interviews were also held at local support organisations. Informal interviews were held with subcontractors and home-workers.
3. This section mainly relied on open interviews with entrepreneurs and at local support organisations, and informal interviews with subcontractors and home-workers.

REFERENCES

Asheim, B. and P. Cooke (1999) "Local Learning and Interactive Innovation Networks in a Global Economy", in E. J. Malecki and P. Oinas, eds., *Making Connections: Technological Learning and Regional Economic Change*, Aldershot: Ashgate, pp. 145–178.

Audretsch, D. B. and M. P. Feldman (1996) "R&D Spillovers and the Geography of Innovation and Production", *American Economic Review* 86(3): 630–640.

Bell, M. and M. Albu (1999) "Knowledge Systems and Technological Dynamism in Industrial Clusters in Developing Countries", *World Development* 27(9): 1715–1734.

Bennel, P. (1999) *Learning to Change: Skills Development among the Economically Vulnerable and Socially Excluded in Developing Countries*, Geneva: International Labour Organization.

Berry, A. and E. Rodriguez (2001) "Dynamics of Small and Medium Enterprises in a Slow-Growth Economy, the Philippines in the 1990s", WBI Working Papers, Washington, D.C.: World Bank Institute.

Caniëls, M. C. J. (1999) "Regional Growth Differentials: The Impact of Locally Bounded Knowledge Spillovers", Ph.D. dissertation, Maastricht: Maastricht University.

Caniëls, M. C. J. and H. A. Romijn (2001) "Small-Industry Clusters, Accumulation of Technological Capabilities and Development: A Conceptual Framework", ECIS Working Paper 01.05, Eindhoven: Eindhoven Centre for Innovation Studies, University of Eindhoven.

Capello, R. (1999) "Spatial Transfer of Knowledge in High Technology Milieux: Learning Versus Collective Learning", *Regional Studies* 33(4): 353–365.

Conceição, P. and M. Heitor (1999) "On the Role of the University in the Knowledge Economy", *Science and Public Policy* 26(1): 37–51.

Florida, R. (1995) "Towards the Learning Region", *Futures* 27(5): 527–536.

Fluitman, F. (1989) *Training for Work in the Informal Sector*, Geneva: International Labour Office.

Gereffi, G. (1999) "International Trade and Industrial Upgrading in the Apparel Commodity Chain", *Journal of International Economics* 48(1): 37–70.

Humphrey, J. and H. Schmitz (1996) "The Triple C Approach to Local Industrial Policy", *World Development* 24(12): 1859–1877.

——— (2000) "Governance and Upgrading: Linking Industrial Cluster and Global Value Chain Research", IDS Working Paper 120, Brighton: Institute of Development Studies, University of Sussex.

ILO (1999) "Job Quality and Small Enterprise Development", SEED Working Paper No. 4, Geneva: International Labour Organization.

Keeble, D. (2000) "Collective Learning Processes in European High Technology Milieux", in D. Keeble and F. Wilkinson, eds., *High-Technology Clusters, Networking and Collective Learning in Europe*, Aldershot: Ashgate, pp. 199–229.

Knorringa, P. (1995) "Economics of Collaboration in Producer–Trader Relations: Transaction Regimes between Market and Hierarchy in the Agra Footwear Cluster, India", Ph.D. dissertation, Amsterdam: Free University of Amsterdam.

—— (2001) "Beyond the Industrial District Model: Urban Cluster Trajectories", in I. Baud, J. Post, L. de Haan and T. Dietz, eds., *Re-aligning Government, Civil Society and the Market: New Challenges in Urban and Regional Development. Essays in Honour of G. A. de Bruijne*, Amsterdam: AGIDS, University of Amsterdam, pp. 245–258.

Lawson, C. (2000) "Collective Learning, System Competences and Epistemically Significant Moments", in D. Keeble and F. Wilkinson, eds., *High-Technology Clusters, Networking and Collective Learning in Europe*, Aldershot: Ashgate.

McFarlane, B. (1998) "The State and Capitalist Development in the Philippines", in K. Sheridan, ed., *Emerging Economic Systems in Asia*, Sydney: Alwin and Unwin, pp. 144–179.

MacKinnon, D., A. Cumbers and K. Chapman (2002) "Learning, Innovation and Regional Development: A Critical Appraisal of Recent Debates", *Progress in Human Geography* 26(3): 293–311.

Malmberg, A. and P. Maskell (2002) "The Elusive Concept of Localisation Economies: Towards a Knowledge-Based Theory of Spatial Clustering", *Environment and Planning A* 34(3): 429–449.

Malmberg, A. and Ö. Sölvell (1997) "Localised Innovation Processes and the Sustainable Competitive Advantage of Firms: A Conceptual Model", in M. Taylor and S. Conti, eds., *Interdependent and Uneven Development. Global-Local Perspectives*, Aldershot: Ashgate, pp. 119–142.

Maskell, P., H. Eskelinen, I. Hannibalsson, A. Malmberg and E. Vatne (1998) *Competitiveness, Localised Learning and Regional Development*, London: Routledge.

Nooteboom, B. (1999) "Innovation, Learning and Industrial Organisation", *Cambridge Journal of Economics* 23(2): 127–150.

Oinas, P. and E. J. Malecki (1999) *Making Connections, Technological Learning and Regional Economic Change*, Aldershot: Ashgate.

Perez-Aleman, P. (2000) "Learning, Adjustment and Economic Development: Transforming Firms, the State and Associations in Chile", *World Development* 28(1): 41–55.

Pietrobelli, P. and T. Barrera (2002) "Enterprise Clusters and Industrial Districts in Colombia's Fashion Sector", *European Planning Studies* 10(5): 541–560.

Pyke, F. and W. Sengenberger (1992) *Industrial Districts and Local Economic Regeneration*, Geneva: International Institute for Labour Studies.

Rabellotti, R. (1999) "Recovery of a Mexican Cluster: Devaluation Bonanza or Collective Efficiency", *World Development* 27(9): 1571–1585.

Sandee, H. (1995) "Innovation Adoption in Rural Industry: Technological Change in Roof Tile Clusters in Central Java, Indonesia", Ph.D. dissertation, Amsterdam: Free University of Amsterdam.

Schmitz, H. (1999) "From Ascribed to Earned Trust in Exporting Clusters", *Journal of International Economics* 48(1): 139–150.
Singh, M. (2000) "Combining Work and Learning in the Informal Economy: Implications for Education, Training and Skills Development", *International Review of Education* 46(6): 599–620.
Storper, M. (1997) *The Regional World: Territorial Development in a Global Economy*, New York: The Guilford Press.

8

Understanding growth dynamism and its constraints in high technology clusters in developing countries: A study of Bangalore, southern India

M. Vijayabaskar and Girija Krishnaswamy

1 Introduction

The rise of information and communication technology (ICT) based sectors, as a lead sector in high-income economies, poses challenges and opportunities for low-income, technologically less equipped countries. The high growth potential of this sector and a relative lack of resources to participate in this sector may lead to a possible aggravation of economic disparities within and between countries (World Bank, 1998). Simultaneously, there is a widespread belief that ICT provides an opportunity for such economies to "leapfrog" and "catch up" with the industrially advanced high-income nations (Perez and Soete, 1988). Since the technologies are new and not yet matured, economies may more easily learn and build up competence in this sector. Further, since important segments of ICT production, such as software, are more skill intensive than capital intensive, economies with a good human capital base are likely to gain from growth of the "knowledge" or "weightless" economy.

Accompanying this process of "informatisation" is a greater integration of domestic input and output markets with the world market, rendering competitiveness in the world market critical for the growth prospects of countries. Building up competence in ICT is therefore tied to its diffusion in other sectors of the economy to improve efficiency and performance. There are, however, few low-income regions that have taken advantage of this "opportunity". India is seen as one such country, having established itself as an important player in the world software services industry (Arora et al., 2001). Though a few other economies such as Taiwan, Brazil, and South Korea have built up comparable technological capability in the electronics industry (Mathews and Cho, 2000; Sridharan, 1996), India has been an exemplar in the software services segment, with a global presence unmatched by most low-income countries. Understanding the sources of such competence is therefore vital, especially that relating to the creation of appropriate organisational structures and processes.

Technology-intensive production tends to concentrate in spatially agglomerated clusters of firms, networked with each other, and with other support institutions such as universities and research institutes, financial firms, etc. (Breschi and Malerba, 2001). The Indian software industry is also primarily concentrated in a few spatial agglomerations in the metropolitan cities, the Bangalore software cluster being the most important. Its significance extends beyond India, drawing comparisons with the Silicon Valley high technology district of the United States (Balasubramanyam and Balasubramaniam, 2000: 350). Home to over 900 software firms consisting of subsidiaries of multinationals, joint ventures, domestic start-ups and large firms, it caters for multiple segments of the software sector. Here, we seek to understand the factors contributing to the growth of the Bangalore software cluster, the prospects for a "high-road" trajectory of sustained innovation, and implications for policy-making.

Existing studies seek to explain the success of the Bangalore software cluster as being the result of government policies in conjunction with its participation in the global value chain on account of its low wage costs (Arora et al., 2001; Heitzman, 1999; Lateef, 1997; Parthasarathy, 2000). Or else studies point to the important role played by multinational corporations (MNCs) in infusing technological dynamism into the sector (Fromhold-Eisebith, 2002; Patibandla and Petersen, 2002). With the exception of Arora et al., Fromhold-Eisebith, and Patibandla and Petersen, these accounts are confined to explaining Bangalore's ability to enter the global value chain and are not concerned with factors conditioning its subsequent growth pattern and possible trajectory. Bresnahan, Gambardella and Saxenian (2001) point to the importance of making this analytical distinction in understanding the future prospects of a cluster. For instance, the pattern of learning that firms go through as a result of their exports may influence the nature of subsequent growth. Also, the studies, with the exception of Fromhold-Eisebith (1999, 2002), take little account of the role played by the organisation of software production prevalent in the cluster. Importantly, none address the larger issue of the ability of low-income economies to "catch up" through the promotion and/or creation of such clusters. The dynamism that clusters evince ought to generate adequate growth and productivity linkages so as to bridge the gap between the high- and low-income economies. Otherwise, the dynamism may only reflect the New International Division of Labour proposition put forward by Frobel, Heinrichs and Kreye (1980).

There is a growing body of literature that seeks to understand cluster-specific factors, especially in knowledge-intensive production, that are deemed critical to a high growth trajectory (Bresnahan, Gambardella and Saxenian, 2001; Coombs et al., 1996; Saxenian, 1994). Here, we address this dimension by comparing characteristics observed in high technology clusters elsewhere with those prevailing in the Bangalore software cluster, and use this exercise to reflect upon the ability of low-income economies to catch up with the advanced capitalist economies. Apart from secondary literature available on Bangalore, the study is based on information drawn from case studies of 20 firms from different segments of the software services segment. In addition, data collected for another study by the first author (Rothboeck, Vijayabaskar and Gayathri, 2001) from a survey of 70 software professionals are used to examine the labour market dynamic, seen by many as the core factor underlying many a high technology cluster.[1] Interviews with key informants, such as industry experts and members of producer associations and financial institutions, are undertaken to cover remaining gaps. The diffusion of "secrets of industry in the air", as is common in industrial districts, renders key informants an important source of information.

This paper is organised as follows. First, based on the literature, we identify key variables that are considered critical to innovation-based high technology cluster development. Next, we delineate the pattern of growth of the Bangalore software cluster. Subsequently we move on to examine the prevalence of relevant institutional variables within the Bangalore software cluster and their contribution to the region's dynamism. Finally, we draw inferences for policy intervention in promotion of IT based clusters.

2 Clustering and knowledge-intensive production

Many studies examine the role played by clustering in sustaining innovation in high technology sectors. The static effects of clustering, such as agglomeration economies and external economies, are distinct from its dynamic effects. Firms in clusters benefit from economies of sourcing inputs and in use of infrastructure such as transport, communication, power, and finance. Agglomeration also improves information flows between producer firms, user firms, and employees, thereby reducing transaction costs and promoting knowledge diffusion. Importantly, clustering enables firms to network with each other, which in turn fosters inter-firm division of labour, with firms specialising in specific stages of the production process and thereby building up competencies in respective processes.[2] In clusters that successfully compete, networking is observed to play a critical role (Humphrey and Schmitz, 2000). Porter (1990) and Best (1991) also stress the importance of a strong interactive relationship between suppliers and firms that may diffuse technical knowledge within the cluster. An interactive relationship between users and producers is also increasingly critical, as mass customisation trends in markets warrant a finer understanding of differentiated customer needs.

Networks in clusters are often not confined to the economic realm. They are embedded in the social and cultural institutions of the region, fostering a high degree of trust that enables firms to minimise transaction costs as well as share information essential to build up innovative capacity. Such factors are especially important in high technology clusters where spill-over effects play an important role in organisational learning. Studies show that *collective learning* is facilitated by interactions between entrepreneurs and inter-firm mobility of employees that diffuse critical tacit knowledge pertaining to new products, processes, or markets (Saxenian, 1994).[3] Such an "innovative milieu" is also supported by technology support institutions such as universities, design centres, and laboratories, apart from local institutions that provide "real services" that enable small firms to access resources that would otherwise be beyond their reach.[4]

It is also argued that in the case of radical innovations,[5] which require a great deal of unlearning for an integrated organisation, loosely knit firms tend to be a more conducive organisational form as "the advantage of non-redundancy is more important" (Nooteboom, 1999: 141). Radical innovations, however, require a threshold level of financial, technical, and organisational resources that would fall far beyond the scope of small firms. In IT based sectors, initial production involves considerable sunk costs in research and development, while subsequent reproduction can be undertaken at almost nil costs. Apart from sunk costs involved in R&D, marketing of the new product or service also involves considerable costs. Hence access to risk capital is essential to firms undertaking innovative activity.

Labour markets also influence the pattern of cluster development. Though theoretically well recognised, this has received little attention in the literature on clustering.[6] A strong labour organisation can resist competing through lowering of wage costs, forcing organisations to compete through innovation, facilitating the cluster moving on a "high-road" trajectory. More importantly, high technology clusters require access to a skilled labour force with the ability to keep abreast of new technologies and contribute to the innovative activity of firms (Bresnahan, Gambardella and Saxenian, 2001). Institutional intervention to ensure continuous reproduction of a skilled workforce therefore constitutes another key factor in sustaining such clusters. In sum, the presence of a strong interfirm division of labour, a cooperative environment, conducive labour markets, and access to technological, financial, and other infrastructural resources are key requisites to sustain innovative dynamism in high technology clusters. Fromhold-Eisebith (2002) interestingly argues that such factors, devoid of external impulses such as that from MNCs, are not adequate to create a dynamic cluster. While it may be true that an isolated cluster may not be able to access frontier technologies, it is more important to understand the organisational features that draw MNCs and enable diffusion of frontier technologies within the cluster. Rather than raise an either/or scenario as Fromhold-Eisebith does, it is important to understand the relationship between MNC affiliates and institutional factors internal to a cluster. Also, as Breschi and Malerba point out, access to a highly skilled transnational workforce is becoming a more important means to infuse innovative dynamism than foreign direct investment by MNCs (Breschi and Malerba, 2001: 822).

Indeed, the importance of these variables may vary according to the nature of specialisation and level of evolution of the cluster. Brusco's typology of clusters with different levels of institutional intervention and support is an attempt to capture this dimension (Brusco, 1992). Thus, interfirm division of labour and networks and/or the nature of institutional support may not be an immediate outcome of clustering, but may result over time due to increases in scale economies and growing technological complexity or growth performance (Breschi and Malerba, 2001: 825). Understanding the evolution of clusters and the accompanying institutional dynamic therefore becomes important.

Despite early realisation of the role that ICT would come to play in the economy, India has been able to gain some leverage in the global market only in the software sector.[7] In terms of initiatives, this can be best attributed to an indirect consequence of state intervention, especially in the initial stages. Scholars contend that policies geared to indigenise software development for imported computing systems and the vacuum created by the forced exit of IBM led to the creation of a team of skilled software developers (Heeks, 1996: 70).[8] Hence, when personal computers enhanced the diffusion of computers and related software development needs, Indian software developers took advantage of the opportunity. This process was also facilitated by policy measures to build a pool of highly skilled labour (Heeks, 1996). This set of initial conditions, aided by the accidental prevalence of English as a medium of communication and training, led to India establishing a first-mover advantage in software development, where a low-cost but relatively skilled workforce is critical. Having said that, we nevertheless need to understand the factors that led to clustering of software development firms in Bangalore rather than other centres in India.

3 Evolution of Bangalore as an industrial district

Ever since the early twentieth century, Bangalore's urbanisation and industrialisation patterns have been strongly state-led. Being a part of the erstwhile princely state of Mysore, it benefited immensely from the industrialisation efforts undertaken by the state, which included establishing a polytechnic and a few state-owned factories. The premier research centre, the Indian Institute of Science, was also set up during this period (Holmstrom, 1998b: 170–171). Subsequently, from the 1950s to the 1970s, its dust-free environment proved conducive for setting up large public sector establishments specialising in sectors such as electronics, aerospace, machine tools, and telephone equipment, apart from a few national defence research laboratories. This spawned a number of upstream and downstream activities, predominantly small and medium ventures that supported these firms.

Though the Indian software industry has its origins in Bombay in the 1970s,[9] shortage of skilled labour and the rising costs of built-in space led to a search for alternative locations. Bangalore, for the reasons mentioned above, together with lower land prices, proved to be a favourable destination. Clustering of electronics-related industries also brought people from all over India to Bangalore, giving the city a very cosmopolitan character and culture, extremely different from most other cities in India. One interviewee suggested that for these reasons Bangalore was the only city in India that most multinational corporation expatriates would be willing to live in.

Such factors, especially the low-cost pool of skilled labour, drew multinational IT firms to Bangalore, Texas Instruments (TI) being the first. TI India was set up in 1986 to "enhance ... presence in the Asia-Pacific region ... India was selected because of its strong educational system in theoretical sciences and engineering, for large technical manpower" (NASSCOM, 1995: 384) and of course for its very large English-speaking labour force. Bangalore was chosen because it was considered to better suit the lifestyles of TI's international staff. The software development centre of TI had a direct satellite link with its headquarters in Dallas, the United States. Excess capacity on that satellite link was to be shared with other businesses that needed such a link. This paved the way for the entry of smaller domestic software companies to undertake data entry and basic software programming jobs for distant clients. In 1989, Hewlett-Packard also set up a fully owned subsidiary in Bangalore that was 100 per cent export oriented. This influx of foreign investment has steadily increased since then, and Bangalore at present is home to a number of wholly owned subsidiaries of other IT related transnational corporations (TNCs) and joint ventures (Balasubramanyam and Balasubramaniam, 2000: 355), apart from a large number of Indian software firms undertaking software development work for multinational client firms. This growth in turn created labour pools with skills that new firms could draw upon.

Quite a few entrepreneurs, in the initial years, were erstwhile employees of overstaffed public sector firms, forced to reduce employment due to policy moves to reduce state support for such ventures. Engineer-entrepreneurs also came from the Indian Institute of Science, the Central Machine Tools Institute and the National Aeronautical Laboratory (Holmstrom, 1998b; Heitzman, 1999).[10] The new liberal policy environment also drew entrepreneurs from among the NRI (non-resident Indian) community wanting to relocate. As Patibandla and Petersen state, "Most of the TNCs' and leading Indian firms' development centers are concentrated in Bangalore. In Bangalore alone there are approximately 140 TNC development centers. There are approximately 750 large and small domestic IT firms. About 40 per cent of India's total exports of $8.3 billion in 2001 have come from Bangalore. There are about 60,000 IT professionals employed in Bangalore" (Patibandla and Petersen, 2002: 23). Kumar points out that, though Bangalore has the second largest number of firms in India at around 20 per cent, this is a bit misleading as many firms headquartered in other cities have their production work undertaken in Bangalore (Kumar, 2001: 4287).

Others, including Lateef (1997), Parthasarathy (2000), Holmstrom (1998b), and Balasubramanyam and Balasubramaniam (2000), have already discussed various aspects of the trajectory described thus far. None of the studies, except that of Fromhold-Eisebith (2002), however, seek to understand the dynamism on the basis of the organisational features of a cluster. Balasubramanyam and Balasubramaniam's study, although it has the explicit aim of studying the nature of the cluster, hardly analyses the effects of agglomeration or the nature of intra-cluster networks or labour markets. Holmstrom's work does attempt to understand Bangalore as an industrial district, but it is confined to a study of electronics and machine tool firms, while Fromhold-Eisebith, although emphasising the role of MNCs in infusing dynamism, does not address the manner in which local cluster-specific institutions interact with MNCs to produce a specific trajectory. As a prelude to our attempt to fill this gap, we provide an overview of the key characteristics of the Indian software industry.

4 The Indian software industry in the global division of labour

The Indian software industry is one of the fastest growing segments in the global software sector, with revenues growing at over 50 per cent per annum since 1996/1997. Though more than 70 per cent of revenues are generated through exports, its share in the world market, at about 1.5 per cent in 1999, continues to be very small (Heeks, 1998; Kumar, 2001). However, it has established itself as a major player in the "services" segment of the industry, accounting for 18.5 per cent of customised software sourced from other nations in 1999 (Kumar, 2001: 4278). That this share has risen from 11.9 per cent in 1991 points to a clear advantage that Indian firms enjoy in this segment.[11]

Exports of software services assume two forms. Either firms send personnel to the client's premises to develop software according to their needs. Termed "body shopping" or on-site work, such services are paid for on the basis of the quantum of labour required/used. Or projects can be on a turnkey basis, where firms are responsible for delivering an entire software package and are paid on a project basis. It is found that clients mostly outsource low-end, labour-intensive tasks to Indian firms, retaining more skill-intensive high-end activities such as design and product development in-house (Heeks, 1998; Parthasarathy, 2000). Low wage costs therefore continue to be the primary competitive factor for Indian firms (Heeks, 1998; Lateef, 1997). Simultaneously, Indian firms have diversified horizontally into various service domains, taking advantage of demand for new types of software services such as Y2K solutions, Euro-conversion, and e-commerce. Among the various domains for which Indian firms undertake software development, banking, insurance, healthcare, retail, transport, telecommunications, and education and training constitute the primary segments (Kumar, 2001: 4282).

In terms of distribution of firms on the basis of size, we observe that a few large firms account for the bulk of the output (*Dataquest*, 15 July 2000). Apart from export firms, firms catering for the domestic market account for 25–30 per cent of the total number of firms in the country. The range of services provided varies across firm size, with the larger ones having a more diversified portfolio compared with smaller firms. A handful of firms, including multinational subsidiaries, are involved in product development as well. However, most firms are involved in development of customised software, rendering networking with client firms an important aspect of the organisation of software production in India. Having highlighted the nature of the market for the Indian software industry and the importance of the Bangalore software cluster, we now move on to evaluate the prospects for Bangalore to evolve into an innovative industrial district. Based on identification of key variables from an earlier discussion, we evaluate their role in the Bangalore cluster.

5 Social division of labour and inter-firm networking

Clustering encourages specialisation of firms in complementary processes and a concomitant build-up of expertise in respective areas, a key feature of capability building in high technology clusters. In the Bangalore software cluster, such a social division of labour is relatively absent. A few big firms accounting for the bulk of the cluster's output dominate the cluster. Such firms undertake production of all the orders they secure in-house. This absence of specialisation has largely to do with the cluster's location in the international division of labour. As pointed out earlier, firms in Bangalore specialise in labour-intensive stages of software development such as low-level design, coding, and testing and maintenance, with client firms supplying them with high-level design after analyses of their requirements. Hence, there is little scope for evolving a fruitful vertical inter-firm division of labour within the cluster. Given that design can be undertaken only after an understanding of user firm needs, it is extremely difficult for firms in Bangalore to move into this activity as user firms are located elsewhere, operating under less understood market conditions. Barriers are also erected by client firms needing to retain the more strategic stages of software development in-house, outsourcing only the routine processes (Arora et al., 2001: 1274).

On the other hand, firms catering for the domestic market do specialise in all segments and some, as has been pointed out by Arora et al. (2001: 1273), have developed extremely complex software system solutions. However, such firms are not linked up to export markets, while firms catering for the export market seldom focus on the domestic market. Fromhold-Eisebith (2002: 2169) argues that the "walking on two legs" strategy of leveraging learning in the domestic market in the global market may not be a feasible one as the market requirements are quite distinct. The contention is true only in a limited sense. As input and output markets are increasingly integrated with global markets, there tends to be a certain degree of convergence of technologies, with the entry of MNCs in other sectors on the one hand, and greater emphasis on exports among domestic firms on the other. An interesting case in this regard is that of I-flex Solutions, a former subsidiary of a multinational bank, which has successfully launched a couple of software products for banking solutions. Based on its expertise developed in-house in Mumbai, it initially catered for low-income country markets, especially in Africa, before moving on to markets in Europe and Canada. Similarly, Infosys Technologies has used its learning acquired in software service exports to launch software products for banking solutions in the domestic market. Though such cases are few, they do point to the importance of developing linkages between exports and production for the domestic market.

Despite barriers to the evolution of a vertical division of labour, the cluster has diversified its output market profile horizontally to cater for a wide range of software services. It might therefore be expected that firms would specialise in specific domains and then network with each other to realise scope economies. On the contrary, evidence of such domain specialisation is once again meagre. Such diversification has been accomplished primarily through vertical integration of various segments within a firm, rather than networking between domain-specialised firms. The nature of the work does not require high-level domain expertise, leaving firms with little incentive to develop any domain-specific specialisation. The rise of large firms employing hundreds of workers, working on various domains, is therefore understandable.

Another reason for the relative absence of firm-level specialisation and networking is the clients' need to ensure the secrecy of codes developed, discouraging outsourcing by Indian vendors. In fact, contracts include "maintenance of secrecy" clauses as an important commitment by vendors. Even the supplier firms resist such outsourcing due to the threat of possible price competition from subcontractors later. The lack of social institutions that bind the firms together and encourage inter-firm exchanges and networking may be another possible factor. Nevertheless, there are exceptions in this regard. To begin with, subsidiaries of MNCs operating in Bangalore do undertake software development exclusively for parent firms, and hence build up domain specialisation. Second, there are also a few firms that do specialise in specific domains such as telecom or finance, as pointed out earlier. Also, new entrepreneurs, to cater for new kinds of services, form specialist firms. Such relatively smaller firms, specialising in domain-specific software production, are however few in number and, more importantly, account for a very small share of the total output (Dossani and Kenney, 2001: 19; Heeks, 1996: 87-90). Also, most MNCs do not have linkages with local firms, though a few of them do enter into collaborative arrangements with training and research institutions.

Networks tend to be established with client firms located elsewhere rather than within the cluster. Since client firms are leading MNCs, vendors are exposed to frontiers in technology use and business practices. Networking with buyers has enabled firms to undertake incremental process innovations. In terms of the software development process, there have been substantial improvements, especially among larger firms. The use of improved software development techniques and an emphasis on quality control are all too visible, even in quantitative terms. Of the 31 firms worldwide that have attained Level 5 of SEI-CMM (Software Engineering Institute's Capability Maturity Model, USA), the topmost level for quality of software development management, 17 are in India, with 13 of them based in Bangalore. A total of 170 Indian firms have secured International Standards Organization 9000 (ISO 9000) certification. Even in labour productivity terms, Kumar finds a slight reversal in the declining trend since 1996/1997, as firms have been able to improve the revenue per unit wage bill (Kumar, 2001: 4281). Though such learning is useful to build up the capability of individual firms, their distant location offers little opportunity to interact and understand user requirements. Diffusion of tacit knowledge, critical to knowledge flows and learning in a high technology cluster, is therefore inadequate.

The importance of the software development process to the growth of the cluster is evident from the fact that the only instance of inter-firm interaction and sharing of expertise in the cluster pertains to this area. In fact, it is said that Bangalore has the largest active software process improvement network (abbreviated to SPIN) in the world. Formed in 1992 and comprising 10 core members drawn from leading software service firms in the cluster, it has become active only since 1999, with regular interactions via email and periodic meetings and talks open to other software developers in the cluster (Hari, 2001).

Interestingly, while ICT is a realm marked by radical innovations and shortened product life cycles, the process of software development as such is relatively less characterised by this rapid technological change. In fact, the inability of software development to match up to productivity increases in hardware or technical changes in telecommunications has led to it being referred to as the "software bottleneck". Despite automation of certain processes and efforts to remove certain stages in software development, the process as such continues to be highly labour intensive, and confers regions with a pool of low-cost skilled labour with a definite advantage. Consequently, firms in Bangalore are less open to threats posed by technological overhauling of processes and products.

Innovations in global software development have been more in the realm of expanding and deepening its application in other sectors. Herein lies the importance of interactions with users. The extensive reliance on the export market denies software firms in the Bangalore cluster such critical interactions. The few firms that have moved into product development and domain specialisation are firms that have been involved in software development for a single client operating in a specific domain over a long period of time. Or else entrepreneurs with previous work experience in MNCs, either in the United States or in India, launch new firms specialising in niche segments. Nevertheless, entry barriers posed by the high sunk costs of marketing and brand building continue to restrict firms from becoming involved in product innovation.

6 Labour markets

Scholars distinguish a "high-road" trajectory, with a well-paid, multi-skilled, "functionally flexible" workforce contributing to innovation-based competition of the cluster, from a "low-road" trajectory characterised by cost-based competition and use of labour market flexibility (Holmstrom, 1998a: 21). A human capital-intensive sector like software development definitely requires the former type of labour force to build up innovative capability. In the case of Bangalore, a set of favourable initial conditions have played an important role, with later developments serving to reinforce such advantages. Bangalore's entry into software exports was primarily due to the large pool of skilled labour created through the numerous high technology public sector firms in the city. The consequent growth in software exports has only reinforced the existing pool of skilled labour, while simultaneously drawing labour from the numerous engineering colleges within the state and elsewhere in southern India. All case firms sourced their labour almost entirely from engineering institutes within southern India with a few exceptions such as the Indian Institutes of Technology and Regional Engineering Colleges. Also, its built environment, highly amenable to the consumption requirements of a blue-collar, well-paid workforce, continues to offer itself as an appropriate place to work, especially among the highly skilled NRIs wanting to relocate in India.[12]

To understand the skill requirements of the cluster, it is imperative to understand the dynamic of technologies of software development. Though the engineering process has not been relatively dynamic, the increasing diversity of applications has led to continuous change in the nature of languages and packages used in software development. For instance, the rise of the internet has led to software tuned for web-based applications. Similarly, the growth in multimedia entertainment has fostered introduction of languages and packages meant for software development in this segment. Continuous learning and an ability to shift from one language/package to another are therefore critical to software development in Bangalore, especially in the context of its specialisation in low-end activities in a diverse spectrum of services.

In terms of access to a workforce amenable to learning and using multiple skills, established export firms do not report many problems, though, by and large, all respondents agree upon the tightness of the software labour market (Rothboeck, Vijayabaskar and Gayathri, 2001).

The recruitment process in big export firms, in fact, prioritises this "learnability" facet of potential entrants above other factors such as existing skill sets. The average age of the workforce at 25–26 years indicates a preference for fresh minds with a perceived ability to learn new skills more quickly (ibid.). In the case of smaller firms, however, we do observe a marked constraint in accessing such a workforce due to their inability to match the bigger firms in salaries offered. On-the-job training, apart from formal training in academic institutions, is important to sustain production. For firms to invest in such training however, the benefits derived from training need to offset the costs incurred. In other words, the cluster as a whole has to effectively utilise the multi-skilled workforce to compete on the basis of innovation. In this regard, we observe serious limitations, even in big firms.

The limitation can be attributed to two primary factors. The global shortage of software labour, especially in the United States, forces firms to source labour from other regions including India. India, with its relatively large pool of labour with a knowledge of English, in fact constitutes the biggest source of external labour for the United States (National Research Council, 2001). The higher salary levels in the sourcing countries prove to be a lure for skilled labour in India, especially Bangalore. As Arora et al. point out, firms in the United States faced with labour shortages draw upon labour from India to work in core areas, outsourcing routine work to software firms in India (Arora et al., 2001: 1276). Second, the inability of firms to specialise in more knowledge-intensive segments of software production like design forces the very skilled to migrate to firms where such activities are undertaken. The two factors, together, pose serious constraints to firms in Bangalore seeking to build up their innovative capability. The movement of personnel from smaller export or domestic firms with lower salary levels to bigger firms prevents the former from accessing a skilled workforce. Simultaneously, the bigger firms lose their more skilled workforce to firms in other countries. Inter-firm mobility in Bangalore, therefore, undermines the ability of firms to upgrade based on the built experience of the workforce. The costs of labour mobility are also high due to investments in training.

The diffusion of "secrets of the industry in the air" in a cluster provides an enabling environment for positive knowledge spillovers. This is definitely true in the case of large firms wherein inter-firm employee mobility has enabled diffusion of skill and expertise. However, given the outward movement of labour, such diffusion of skills takes place only at the lower end of the spectrum. Such spillovers therefore do not translate into benefits to firms within the cluster. Labour turnover therefore constitutes a major hurdle to technological upgrading in the cluster.[13] Diffusion of skills has nevertheless resulted in the formation of new firms, especially ones that offer new kinds of services, and, more importantly, firms that offer standardised packaged software, at times even for the domestic market. The relocation of many non-resident Indian professionals has also led to many of them establishing high technology ventures, especially in software (Khadria, 1999). Thus, while the cluster has definitely been constrained due to out-migration of the workforce, the NRI relocation back to Bangalore partly offsets the effects of labour poaching.

Of late, a contraction in demand due to recession in some of its biggest markets, especially the United States, has led to employment flexibility too being used. Informants reveal that, while smaller firms witness higher firm mortality and consequent job losses, and medium-sized firms witness lay-offs, the bigger firms are experiencing a reduction/freeze in recruitment accompanied by cuts in fringe benefits, etc. (interview with Ramadoss by author, key informant, Bangalore, 28 August 2001). Such employment flexibility, together with the inability of the cluster to access a stable, multi-skilled workforce for innovation, indicates tendencies to a low-road trajectory, partly offset by the rise of new service firms on account of labour circulation and consequent entrepreneurship. Having highlighted the nature of inter-firm networks and labour markets in the Bangalore software cluster, we now move on to examine the other important element of an innovative cluster, the provisioning of critical services essential for build-up of the innovative capability of the cluster.

7 Real services

7.1 Infrastructure

Provisioning of services requires, to begin with, appropriate institutional mechanisms to identify the needs of a cluster, followed by measures to provide the required services.[14] We have already highlighted the role of public sector enterprises in drawing software capital to the cluster, through creation of a pool of skilled labour. Another critical input to software service production is the availability of communication and hardware support. In both respects, the state has had a major role to play. At the state level, policies fostered private participation in higher education, leading to the establishment of a number of engineering colleges, thereby enhancing the supply base for labour. Subsequent to the software boom, at the national level, private sector participation in software training is encouraged and, at present, training is an important segment of the software sector in India (Kumar, 2001).

With regard to physical infrastructure, in the mid-1980s the Karnataka state government in conjunction with the Department of Electronics

(DoE) created an Electronics City. It provided the facilities and infrastructure necessary to promote investment in the electronics industry, including a guaranteed supply of electricity, telecommunications facilities, and a technical training centre. Financial incentives such as cheaper credit and tax relief were also offered.[15] They were later extended to software when it became apparent that this would be an important source of revenue, enabling firms to undertake offshore projects. While earlier, on-site services were composed nearly 100 per cent of exports, with the establishment of software technology parks and accompanying data communication facilities there has been a steady increase in offshore software development, which at present accounts for more than 50 per cent of exports (Parthasarathy, 2000). This movement has definitely helped firms to move into more complex projects and compete at lower prices, and importantly provides the opportunity to create linkage effects in the cluster apart from enhancing net foreign exchange earnings.

In addition, the state government, in collaboration with a Singapore-based firm, has built a technology park, which provides excellent data communication and power facilities for software and other high technology firms. While hardware imports also have been liberalised, efforts to develop a vibrant hardware sector have been negligible. In terms of communications infrastructure, Bangalore is well connected to other cities in the world through a gateway located in the city, with one of the highest bandwidths in India (http://www.bangaloreit.com/html/itscbng/itinfra.html). It is also connected to other cities in India through wideband optical fibre and microwave media, though further improvements may be required with the growth in e-commerce related ventures. The next most important requirement in a high technology cluster is access to risk capital for innovative ventures.

7.2 Finance

The cluster has only recently begun to need innovative financial instruments. Onsite services did not call for much capital investment as it was largely export of labour. Only with the movement to offshore projects did the need for physical infrastructure such as computers and communication equipment arise. And since in most cases firms moved into offshore projects from body shopping, they had surplus capital to invest in infrastructure. However, new firms that sought to move directly into offshore projects needed investment assistance. With the growth in offshore processing and attempts by a few firms to enter product development, venture capital has become important.

Conventional sources like bank loans were difficult to access given the lack of collateral for these firms. Further, the relatively high interest rates

Table 8.1 VC/angel investments in high tech firms in India

Year	Rs million	US$ million
1996	700	20
1997	3,200	80
1998	6,100	150
1999	14,000	320
2000[a]	32,000	750
2001[a]	50,000	1,200
2008[a]	450,000	10,000

Source: http://www.american.edu/carmel/ph0616a/page13.html.
a. Projected figures.

prevailing in the economy made traditional means of borrowing inefficient, especially when firms compete internationally to access markets. It is in this context that venture capital (VC) firms began to gain prominence in the Indian software sector. The government's role has been twofold. On the one hand, in active collaboration with industry experts, it has sought to facilitate the entry of VC firms into the country. On the other hand, government financial institutions and even state industry promotion agencies, including that in Karnataka, began to set up VC arms of their divisions to promote entrepreneurship. In addition, nationalised banks also began to move into venture funding. More importantly, the last few years have witnessed the entry of foreign institutional investors into this segment, which at present constitutes the largest share of VC funding in India (Dossani and Kenney, 2001: 31). Together, these factors have led to a sharp rise in the growth of VC flows into high technology sectors including software (table 8.1).

Bangalore has been a beneficiary of all these initiatives. In fact, the first venture capital institution in India, the Technology Development and Information Co. of India Ltd. (TDICI), a subsidiary of the state-owned and state-run financial institution ICICI, was established in Bangalore in 1988 (Dossani and Kenney, 2001: 23). TDICI funded many a venture in the early phases, including success stories such as Microland, VXL, Mastek Software Systems, and even a project for Wipro (ibid., p. 25). Though Bangalore ranks only third in the number of venture capital firms locating their offices in India, it hosts the bigger venture capital firms such as Draper International, TDICI and others (ibid., p. 32). Further, Walden International, a leading venture capital firm, estimates that Bangalore accounts for the largest share (35 per cent) of the total risk capital flowing into the country (Hari, 2001). To identify the importance of venture capital for the growth of the industry, we undertook a case study of a successful firm that had relied on venture capital, and gathered information from key informants.

Funded by an international VC firm, the funds were essentially for services. The entrepreneur, with a long stint in a multinational software firm and known in industry circles, had no difficulties in getting funds. Interestingly, at that stage, the VC firm even drew up the business plan, and all that the entrepreneur had to do was to recruit the right people and coordinate production. As the VC firm had a global clientele of over 500 firms, it was easy for this start-up to procure orders from clients. The VC firm's reputation, aided by introduction to its clientele, enhanced access to markets. To that extent, there existed a synergy between the two firms' strengths. Once established, the firm decided to launch a product for which it sought funding from another VC company, once again a multinational (MN) firm. A new firm was developed which concentrated solely on the sale of the product developed. The reason for opting for a new VC was the new clientele it would bring. The respondent says that since the VC firm knew of the success of the earlier venture, and given the high demand for such a product at that time, there was no hitch in accessing funds.

As most firms specialise in low-end services, VC is required more for marketing and accessing clients and to an extent for scaling up of current operations, rather than for product innovation in Bangalore. Respondents suggest that they prefer funding from reputable multinational firms, as, apart from drawing upon a premium clientele, it helps firms to use the former's reputation as an indication of their firm's ability to deliver. And precisely for the same reason, they seldom approach VC firms set up by government institutions to fund their requirements. In fact, sourcing from the latter may diminish their reputation among foreign clients as it may be seen to reflect their inability to source from premiere MN VC firms with good technology valuation procedures. Further, it is said that the ceiling on venture funding from state institutions is much lower than that allowed by foreign VC firms. However, state-supported VC institutions are agreed to have a role to play in the development of software for the domestic market.

The Government of India has also nucleated a National Venture Capital Fund for the Software and IT Industry (NFSIT). Set up in association with various financial institutions and the industry (http://www.american.edu/carmel/ph0616a/page13.html), it encourages entrepreneurship in the areas of software, services, e-commerce, and other IT-related segments in which India has built up competence. Nevertheless, respondent entrepreneurs contend that it is difficult for entrepreneurs with truly innovative ideas or products to acquire seed capital, as there are hardly any entrepreneurial networks within the cluster, unlike the situation in successful clusters such as Silicon Valley where substantial venture funding was through high technology-based entrepreneurs themselves. In India, most VC funds are from financial institutions less willing to undertake such innovative and risky ventures primarily because of India's location in the global division of labour, which poses barriers to moving into more innovative segments. As Arora et al. point out, "Firms that are trying to develop software products have faced problems in getting finance, in part because of the inexperience and conservatism of Indian venture capital funds" (Arora et al., 2001: 1281). Rather, funds have largely gone into funding service firms that undertake more routine software development work with almost assured markets or, of late, to IT-enabled services that are low-end and involve much less risk. Though such lack of funding may not affect the prospects of firms in the cluster at present, it would definitely pose constraints to moving onto a high-road path. Finally, exit options were considered to be few, with the general feeling that entrepreneurs were unwilling to sell their start-ups even if it were feasible.

7.3 Technology and marketing support services

Throughout the 1990s, the state government has been quite successful in marketing Bangalore as a centre for foreign investment in the ICT segment. It has supported initiatives such as collaboration between the European Commission (EC) and the Department of Electronics of the Government of India in setting up the Software, Services, Support and Education Centre Limited (3SE). The purpose of 3SE is to "promote cooperation between the EC and India in the field of computer software" (3SE brochure). It recognises that the Indian software industry needs to diversify its markets and that European companies have been slower to take advantage of the high-quality/low-cost software development environment that India provides, in terms of their sourcing activities. 3SE provides information services about Indian software companies to European companies looking to outsource or form alliances, in terms of matching needs. It also informs Indian software users of products brought out by EU companies.

Apart from such support in marketing, a few institutions have been established to provide technology and training support to the industry. The Indian Institute of Information Technology, Bangalore, and the Institute of Bio-Informatics are novel experiments in public–private partnership that seek not only to train high-quality professionals for the software industry, but also to encourage interaction between academia and experts from industry.[16] In addition, the Software Engineering Institute has been set up as a joint venture between the state government, the Indian Institute of Science, LG Soft India and the Center for Information Systems Engineering of the Carnegie Mellon Research Institute for Information Systems Engineering. This institute also offers advanced training in software engineering (Patibandla and Petersen, 2001: 14; 2002: 1572–1573). Interaction between academia and industry, which has always been low due to earlier governmental restrictions on university employees participating in commercial ventures, has increased in recent years, especially with the formation of the Society for Innovation and Development by the Indian Institute of Science (IISc) in the late 1990s. Since then, the number of collaborative projects with the industry has increased to 80 from 10 only 3 or 4 years ago (Hari, 2001). The Institute has also been able to incubate a few high technology firms of late.[17] The development of the "Simputer" by a set of scientists at the IISc is a case in point. This is purported to be a low-cost alternative to the personal computer that can be shared by many users and allows user interfaces based on sight, touch, and sound. It is hoped to find diverse applications for it in rural India where literacy levels are low. The Simputer Trust consists of both academics and members from the industry.

Simultaneously, a number of MNCs have established research laboratories and development centres focusing on frontier technologies to take advantage of the low-cost highly skilled labour force in the cluster. This will in turn create a pool of skilled labour that can be tapped into by future start-up firms. Hari (2001) cites the case of a high technology venture which recruited its first 20 technical personnel from within the cluster itself. More definite understanding of the impacts of such foreign investments needs a detailed study. The state, as can be seen, nevertheless continues to play an important role in ensuring that at least a certain number of important real services are made available to firms in the cluster. While earlier the role of the state was rather indirect, along with the growth in exports there has been a concerted effort to improve the competitiveness of the cluster by both state institutions and/or producer associations.

8 Implications for policy intervention

The growth performance of the Bangalore software cluster is not easily observed in clusters located in other low-income regions. The establishment of strong networks with clients, investments by numerous subsidiaries of MNCs, return migration of skilled labour, and the resultant entrepreneurship in high technology segments have led to diffusion of skills and knowledge of frontier technologies in software production. This, coupled with competitive pressures, has led to improvements in processes, emphasis on quality control, improved employee productivity, and limited movement into more value-adding services. The state too has played an important role in ensuring a steady supply of skilled labour, apart from positive interventions in the realm of infrastructure and technology support services. In terms of access to finance, few existing firms studied reveal constraints. Despite the presence of factors recognised as contributing to an innovative milieu, the primary basis of competition continues to be one of low wages and a large pool of skilled labour. Attempts to move beyond such low-road competition have not met with much success so far. Further, we observe serious differences in firms' ability to access skilled labour even within the cluster, with the smaller firms and firms catering for the domestic market unable to compete with established well-paying large export firms.

The most important constraint facing the cluster is a relative absence of inter-firm division of labour and networking, especially with regard to user–producer interaction. The excessive focus on the export market prevents firms from interacting closely with users so as to understand the latter's needs better, which is becoming increasingly critical to software development. While firms have definitely built up adequate competence in software development, an inability to move into design prevents them from building up further capability in this sector. The out-migration of skilled and experienced labour compounds this inability, as it tends to push firms further away from investing in training and competing on the basis of innovation. Even with regard to financing, though constraints on venture capital flows have eased considerably, such flows tend to be directed mostly to less risky, relatively low-end software development ventures. Funds for product innovation have hardly found their way into the cluster. This is due not only to a lack of inter-firm networking within the cluster, but also to constraints faced by virtue of its positioning in the global division of labour as a cluster providing complementary services to technology leaders in high-income countries. A lack of intra-cluster division of labour and specialisation in low-end segments of the vertical chain also retard individual firms' tendency to acquire domain-specific competence as observed in more innovative clusters.

State policies need to be directed towards promotion of the domestic market and leveraging of this learning in the global market. Provision of tax and credit incentives to export firms and a consequent reduction in incentives to produce for the domestic market may undermine the scope for such learning. To overcome these limitations, not only is state intervention critical, but so is the "strategic intent" of lead firms in the cluster. Given that a few big firms dominate the cluster, it is imperative that these firms direct the activities of the cluster in such a way as to strengthen its innovative capability. Entering into strategic alliances with MNCs in areas such as brand building and marketing may prove to be increasingly critical. Simultaneously, the firms need to work in close collaboration with state institutions to ensure that critical resources are provided. Limited ability to retain skilled labour and the relative lack of inter-firm division of labour and networking with users may, however, continue to pose constraints to transition to a "high-road" trajectory.

Two strategies are suggested for nascent clusters to link up to markets. One involves specialising in technologies complementary to technology leaders like the Bangalore cluster and then striving to "move up the value chain". Constraints on such a strategy are evident from the above discussion. The other strategy is to specialise in segments not covered by the technology leaders, i.e. to "run a different race" rather than seek to "catch up". The importance of the domestic market, especially large ones such as those of India or China, to the latter strategy cannot be overstated. Importantly, failure to leverage learning from exports to cater for the domestic market throws up the larger issue of the ability of economies to "catch up". If the knowledge generated serves more to enhance the productivity of firms in high-income economies, with few spillovers into the rest of the host economy, the dynamism may only reinforce existing global divisions of labour. Herein lies the importance of strong institutional intervention to ensure that the innovative capabilities of a cluster are directed to create positive linkages in the local economy.

Notes

A draft version of this paper was presented at the 10th Annual Conference of the European Association of Development Research and Training Institutes (EADI) held at Ljubljana, Slovenia, during September 19–21, 2002. The authors are thankful to K. Vaijayanti for assistance provided during fieldwork. Thanks are also due to Mr. Ashok Taluqdar (Cellnext, Bangalore), Mr. Milind Priolkar (BedRoq, Bangalore), Prof. S. Sadagopan and numerous other key informants for critical support, without which the study could not have been completed. Useful inputs from Dr. Henny Romijn, Dr. Sunil Mani and participants in the above-mentioned conference are gratefully acknowledged.

1. "The mobility of skilled workers represents, in fact, the crucial source of new firm formation as well as the main mechanism through which technical and market knowledge flows locally" (Breschi and Malerba, 2001: 821).
2. "For a firm to increase or deploy its own knowledge effectively, it may have to complement this knowledge with that of other firms; and more often than not, by way of some kind of collaborative agreement" (Dunning, 2000: 10).
3. We use the term "collective learning" as opposed to the rather static term "collective efficiency" to capture the importance of the dynamic learning process to a high technology cluster.
4. The notion of real services was introduced by Brusco (1992), by which he means services provided by local organisations in real terms as opposed to provision of financial assistance to firms to acquire the same. Information on technological change, raw material prices, market trends, and equipment testing facilities are examples of "real services" provided in dynamic clusters.

5. Innovations that involve a quantum leap in the production frontier or creation of new frontiers.
6. Two works that explicitly comment on the role of labour organisations in influencing the source of competitiveness of industrial clusters are Best (1991) and Schmitz and Musyck (1994).
7. The different trajectories that different segments of ICT, such as telecoms and software, have assumed and the role of government policy in this regard have been discussed by Parthasarathy (2000).
8. In fact, quite a few ex-IBM employees moved on to establish small software firms (Parthasarathy, 2000).
9. The first software export park was set up there in 1991.
10. "The public sector was, in this sense, a stage in the assembly of capital and human resources that established a critical mass of market opportunities and people, allowing the emergence of an internationally competitive high technology" (http://www.epw.org.in/showArticles.php?root=1999&leaf=01&filename=2935&filetype=html#top).
11. Customised software is developed for specific clients as opposed to generic software packages or products that are sold in the market for wider, mass consumption.
12. A study of Indian emigrant labour in the Silicon Valley, USA, by Dossani reveals that though in terms of origins of the workforce Bangalore ranks fifth, it is first as the preferred destination for relocation of these employees in India (Dossani, 2002: 7, 18).
13. Of course, this is not to say that labour market constraints are the only constraints on a high-road trajectory in Bangalore.
14. The role of policy measures has been described by many, including Bajpai and Shastri (1998).
15. See http://www.dqindia.com/content/search/showarticle.asp?artid=21969.
16. The former has an incubation centre that has already enabled two start-up firms.
17. In addition, this has also facilitated some faculty members of IISc to invest in the equity of a bio-informatics firm (Hari, 2001).

REFERENCES

Arora, A., V. S. Arunachalam, J. Asundi and R. Fernandes (2001) "The Indian Software Services Industry", *Research Policy* 30(8): 1267–1287.
Bajpai, N. and V. Shastri (1998) "Software Industry in India: A Case Study", Harvard Institute of International Development, Development Discussion Paper No. 667, Harvard: Harvard University.
Balasubramanyam, V. N. and A. Balasubramaniam (2000) "The Software Cluster in Bangalore", in J. H. Dunning, ed., *Regions, Globalization, and the Knowledge-Based Economy*, Oxford: Oxford University Press, pp. 349–363.
Best, M. (1991) *The New Competition: Institutions of Industrial Restructuring*, Harvard: Polity Press and Harvard University Press.
Breschi, S. and F. Malerba (2001) "The Geography of Innovation and Economic Clustering: Some Introductory Notes", *Industrial and Corporate Change* 10(4): 817–834.
Bresnahan, T., A. Gambardella and A. Saxenian (2001) "'Old Economy' Inputs for 'New Economy' Outcomes: Cluster Formation in the New Silicon Valleys", *Industrial and Corporate Change* 10(4): 835–860.

Brusco, S. (1990). "The Idea of an Industrial District: Its Genesis", in F. Pyke, G. Becattini and W. Sengenberger, eds., *Industrial Districts and Inter-firm Co-operation in Italy*, Geneva: International Labour Organization, pp. 10–19.

Coombs, R., A. Richards, P. Saviotti and V. Walsh, eds. (1996) *Technological Collaboration*, Cheltenham: Edward Elgar.

Dossani, R. (2002) "Chinese and Indian Engineers and Their Networks in Silicon Valley", draft paper, Stanford, Calif.: Asia-Pacific Research Center, Stanford University.

Dossani, R. and M. Kenney (2001) "Creating an Environment: Developing Venture Capital in India", BRIE Working Paper No. 143, Berkeley, Calif.: University of California, online at: http://repositories.cdlib.org/brie/BRIEWP143.

Dunning, J. H. (2000) "Regions, Globalization, and the Knowledge-Based Economy: The Issues Stated", in J. H. Dunning, ed., *Regions, Globalization, and the Knowledge-Based Economy*, Oxford: Oxford University Press, pp. 7–14.

Frobel, F., J. Heinrichs and O. Kreye (1980) *The New International Division of Labour*, Cambridge: Cambridge University Press.

Fromhold-Eisebith, M. (1999) "Bangalore: A Network Model for Innovation-Oriented Regional Development in NICs?", in E. J. Malecki and P. Oinas, eds., *Making Connections. Technological Learning and Regional Economic Change*, Aldershot: Ashgate, pp. 231–260.

—— (2002) "Regional Cycles of Learning: Foreign Multinationals as Agents of Technological Upgrading in Less Developed Countries", *Environment and Planning A* 34(12): 2155–2173.

Hari, P. (2001) "Bangalore: Technopolis", *Business Standard*, 26 February 2001, online at: http://www.firstandsecond.com/bw/mag2_page3.html.

Heeks, R. (1996) *India's Software Industry: State Policy, Liberalization and Industrial Development*, New Delhi: Sage.

—— (1998) "The Uneven Profile of Indian Software Exports", Development Informatics Working Paper Series, Working Paper No. 3, Manchester: Institute for Development Policy and Management, University of Manchester.

Heitzman, J. (1999) "Corporate Strategy and Planning in the Science City: Bangalore as 'Silicon Valley'", *Economic and Political Weekly* 34(5): PE2–PE11.

Holmstrom, M. (1998a) "Introduction: Industrial Districts and Flexible Specialization – The Outlook for Smaller Firms in India", in P. Cadenne and M. Holmstrom, eds., *Decentralized Production in India: Industrial Districts, Flexible Specialization, and Employment*, New Delhi: Sage, pp. 7–44.

—— (1998b) "Bangalore as an Industrial District: Flexible Specialization in a Labour-Surplus Economy", in P. Cadenne and M. Holmstrom, eds., *Decentralized Production in India: Industrial Districts, Flexible Specialization, and Employment*, New Delhi: Sage, pp. 169–229.

Humphrey, J. and H. Schmitz (2000) "Governance and Upgrading: Linking Industrial Cluster and Value Chain Research", IDS Working Paper No. 120, Brighton: Institute of Development Studies, University of Sussex.

Khadria, B. (1999) *The Migration of Knowledge Workers: Second Generation Effects of India's Brain Drain*, New Delhi: Sage.

Kumar, N. (2001) "Indian Software Industry Development: International and National Perspective", *Economic and Political Weekly* 36(45): 4278–4290.

Lateef, A. (1997) "Linking up with the Global Economy: A Case Study of the Bangalore Software Industry", ILO Working Paper, Geneva: International Labour Organization.

Mathews, J. A. and D. S. Cho (2000) *Tiger Technology: The Creation of a Semiconductor Industry in East Asia*, Cambridge: Cambridge University Press.

NASSCOM (1995) *Indian Software Directory 1995–1996: A Compilation of Software Expertise of NASSCOM Members*, New Delhi: NASSCOM.

Nooteboom, B. (1999) "Learning, Innovation and Industrial Organisation", *Cambridge Journal of Economics* 23(2): 127–150.

National Research Council (2001) *Building a Workforce for the Information Economy*, Washington, D.C.: National Academy Press.

Parthasarathy, B. (2000) "Globalization and Agglomeration in Newly Industrializing Countries: The State and the Information Technology Industry in Bangalore, India", Ph.D. dissertation, Berkeley, Calif.: University of California.

Patibandla, M. and B. Petersen (2001) "Role of Transnational Corporations in the Evolution of a High-Tech Industry: The Case of India's Software Industry", Working Paper No. 5, Copenhagen Business School, Denmark.

——— (2002) "Role of Transnational Corporations in the Evolution of a High-Tech Industry: The Case of India's Software Industry", *World Development* 30(9): 1561–1577.

Perez, C. and L. Soete (1988) "Catching up in Technology: Entry Barriers and Windows of Opportunity", in G. Dosi, C. Freeman, R. Nelson, G. Silverberg and L. Soete, eds., *Technical Change and Economic Theory*, London: Pinter, pp. 458–479.

Porter, M. (1990) *The Competitive Advantage of Nations*, London: Macmillan.

Rothboeck, S., M. Vijayabaskar and V. Gayathri (2001) "Labour in the New Economy: Case of the Indian Software Labour Market", monograph, New Delhi: International Labour Organization.

Saxenian, A. L. (1994) *Regional Advantage: Culture and Competition in Silicon Valley and Route 128*, Cambridge, Mass.: Harvard University Press.

Schmitz, H. and B. Musyck (1994) "Industrial Districts in Europe: Policy Lessons for Developing Countries", *World Development* 22(6): 889–910.

Sridharan, E. (1996) *The Political Economy of Industrial Promotion: Indian, Brazilian, and Korean Electronics in Comparative Perspective 1969–1994*, Westport, Conn.: Praeger.

World Bank (1998) *World Development Report 1998–99: Knowledge for Development*, Washington, D.C.: The World Bank.

9

Culture, innovation, and economic development: The case of the South Indian ICT clusters

Florian Arun Taeube

1 Introduction

How can the Indian success in information and communication technologies (ICT) be explained? Is it a result of the liberalisation of the 1990s? If so, then why have other sectors not produced such an impressive performance? Or is it rather influenced by other factors, which have been released – or rediscovered – through this liberalisation, for example a culture of entrepreneurship and innovation? Or may there even be a predisposition towards the so-called knowledge economy?

This chapter is concerned with the successful development of the software industry as the major component of ICT in India. It is inspired by two noteworthy phenomena concerning this industry. First, most of the studies analysing the Indian software industry cover basically the major centres in Bangalore, Hyderabad, and Chennai, all South Indian. Secondly, while there has always been a tradition of entrepreneurship by a particular social group, the merchant and trader caste of the *Vaishyas*, the software industry witnesses more than a proportionate presence of *Brahmins*, traditionally the priestly and knowledgeable caste group.

Thus the central questions are why do Brahmins participate to such an extent in the new industries and, subsequently, why are they concentrated mainly in the south of the subcontinent? Consequently, culture will be examined through the variables caste and ethnicity. It is important to emphasise that these variables cannot be more than complementary to other explanations of the success of the Indian IT industry. Basically, other contributing factors, such as R&D and technology policy, are omitted from the analysis, for there already exists a large body of research on these systemic factors.

Therefore, this chapter employs a mixed methodology. First, an attempt will be made to make the notion "culture" workable in terms of a useful definition. This is done by examining the relevant literature. Later this definition will be applied to the Indian context through identification of characteristic features of Indian society and the Indian economy. Second, this application then leads in to the empirical section in which I have tried to analyse the societal composition of the Indian entrepreneurial "class" of the so-called knowledge economy. This preliminary analysis is based on a small set of existing interview-based studies in which the interviewees' names are taken as an approximation of their cultural background in terms of societal affiliation or, more precisely, social and regional origin. While the use of interview data has its limitations, especially when there is only a limited amount of data available, it opens new directions for further research and should, therefore, be seen as a basis for more empirical work.

The chapter is structured as follows. Section 2 examines the large field of anthropological and social science work to produce a definition of culture which is useful in the economic context. Section 3 takes a closer look at the Indian and, in particular, the South Indian situation. Section 4 then applies the resulting hypothesis to the Indian software industry. Section 5 concludes.

2 Economic culture and development

There is a renewal of interest in the relationship between culture and (economic) development. This is manifested in the debate on "Asian values" as an explanation for the success of the so-called tiger economies. Having started their catch-up from a level similar to other developing countries, their much more successful results are often imputed to their favourable culture. Another recent example is the collective volume edited by Harrison and Huntington (2000) which brings together scholars from different social sciences.

Why can it make sense to include culture in an economic analysis? How can this variable be integrated? Where are the shortcomings of such an approach? A simple but somewhat naïve answer to the first question is to single out which of the characteristics different cultures possess are most likely to result in socio-economic progress in order to replicate these positive attributes in other cultural settings. The challenge of approaches which stress the importance of culture and other social factors that are non-economic in nature but influence the economy and are influenced by it is not merely to impute differences in economic performance to the cultural factors, as earlier scholars often did, but to integrate culture as a variable into theoretical models (Klump, 1996). An interesting approach has recently been taken by Frederking (2002), who tries to carve out the substance of the relationship between culture and development in order to achieve a basis for cross-cultural comparisons.

For this purpose, a quite generic approximation of culture has to be found and applied. Later in this chapter, I will use the variables caste and ethnicity. This is a customisation to the object of study, but could reasonably be part of a broader aggregate of cultural factors. And, although it is not primarily an economic one, focusing on the economic implications of these proxies narrows the scope of this chapter.

At the outset, one has to emphasise how difficult it is to consistently integrate the mutual interdependences that exist with regard to culture (Klump, 1996). Basically, there are two dimensions in this relationship. First, culture can be seen as an end in itself, that is, a good which should be preserved as part of a larger spectrum of goals that should be reached in the course of development. Second, it is also a means to development, both directly through cultural investments and, more importantly, indirectly through values and norms working in a society (Sen, 2001). Since the majority of development theorists see culture rather as a means than as an immediate goal, at least in the near future, I will concentrate on this. Then the question is how culture exerts its influence on the economic realm (Ruttan, 1988). First, there are the often-cited values, beliefs, traditions, and norms. However, they are rarely observable, and hence difficult to measure. Therefore the more visible manifestations of culture are actions, behaviour, and actual social practices which are usually influenced through norms and values.

Focusing on the relationship between culture and the economy and recognising the fact that there are manifold problems with the definition of culture (Gupta, 1994),[1] in order to make it operational in the context of economic development the analysis should be narrowed to economic culture. What, then, is economic culture? Certainly, it is part of the larger cultural setting of a society. The methodological problem that integrating culture with economics poses to the economist is that it is difficult to define and quantify separately from other factors, for example institutional ones. The economic sociologist DiMaggio (1993: 27) says "aspects of culture shape economic institutions and affairs ... economic processes have an irreducible cultural component". In the same vein is the statement by North (1990: 37) that culture underpins the "rules of the game" in any society and provides "the informal constraints on human interaction". Instances of the economic culture approach are the concepts of social capital (for a review see Fukuyama, 2000) and social capability (Abramovitz, 1986). The first approach stresses the *informal* values and norms shared by a small community allowing for trust and cooperation. The latter is more macro oriented and focuses on the ability of a *country* to innovate and adapt to changing external factors in order "to exploit emerging technological opportunities" (Abramovitz, 1986: 406). It stresses the interplay of different factors ranging from education, institution, and openness in enabling a country to adopt new technologies and is related to the literature on systems of innovation (Lundvall, 1992; Nelson, 1993), which vary, however, in terms of geographical focus, i.e. their perspective stretches from national to regional and local. The focus of the innovation systems literature is certainly not on culture, but it can be a complementing factor. The literature dealing with the culture of economic development demonstrates predominantly one line of argument. Most of the authors working on economic culture try to single out those cultural factors that can be positively correlated with development in the sense of modernisation and growth (Lal, 1998b). Usually mentioned are, for example, trust, rationality, the value of work, and religion (Grondona, 2000).[2] This reasoning can be traced back at least to Max Weber's ideal types and has its latest revival in the "Asian values" debate, which has been dubbed a "neat reversal of Max Weber's famous thesis" (Lal, 1998a: 2). These values characterising the highly hierarchical societies of East Asia, however, do not really match the *South* Asian context. As Sen (1999) points out, there are ancient Indian traditions and values that are in contrast to those of the Sinitic societies to the east. An analysis of the principles of the ancient Indian thinker Kautilya shows that they are more egalitarian and condemn such authoritarian approaches as those of East Asia (Sarkar, 2000; Sen, 1999).

However interesting the value debate, culture is manifested and observable only though action or social practices. In summary, culture can be defined as a "socially transmitted heritage", which opens the door for analysis of actual behaviour. The problem with interpreting cultural influences through the study of literature, or the "book view", is that people often act differently from the way in which they would prefer to behave. An instance of such a "preference falsification", it is argued, might be at work in the caste system (Kuran, 1987). This leads me to the Indian situation.

3 South Indian economic culture

This section is a brief theoretical attempt to outline the basic features of an economic culture that can be derived from a Brahmin and South Indian background. The question is whether there exists a regional culture of innovation resembling that of Silicon Valley (Gertler, 1997; Saxenian, 1994). Therefore, Indian culture has to be analysed from a disaggregate perspective.

Broadly speaking, for Indian society the cultural framework is "Hinduism", which provides a comprehensive philosophy rather than "merely" a religion (Dehejia and Dehejia, 1993). But what is commonly known as belonging to "Hinduism" is only part of the more complex Hindu civilisation – it is embedded in an all-encompassing worldview (see, for example, Rothermund, 1995; Stietencron, 1995). Albeit the differences within that composite of religious beliefs are too subtle to be explored here, the broader cultural view allows for the observation of several characteristics.

According to the distinction made in the section above, I will first deal with the issue of values and then proceed to that of observable behaviour. The section ends with an analysis of South India.

3.1 Values and caste

Is there anything that makes the Indian economy an unusual object of analysis, something that precludes a conventional economic study? Presumably there is, some would argue, given the unique phenomenon of caste as the characteristic feature of Indian society. On the other hand, there are those who downplay the influence of caste and, moreover, the institution as such as an invention of "orientalist" scholars. Their most compelling argument is that some social structure similar to the caste system existed long before the arrival of the British, but, through their desire to rationally understand Indian society, they hierarchically institutionalised the rather informal norms with the support of Indian elites (Bayly, 1999). While the issue whether there is a system of caste is heavily disputed in the anthropological literature, it is quite safe to assume a certain influence of caste. That is why it is taken as one of the proxies for culture in this analysis.

However, the term caste is used in two different contexts. First, it is used to describe the *jati*, hierarchically ranked endogamous kinship groups with a regional base centring around the performance of traditional occupations (such as leatherworkers, priests, merchants, or tailors) in an interdependent relationship with other jatis. Secondly, it depicts the more aggregate societal structure of a class-like division, the *varnas* (Bayly, 1999).

Describing the caste system as consisting of a fourfold hierarchy of varnas plus the so-called untouchables, mistakenly referred to as outcastes (Dumont, 1980),[3] is a gross oversimplification (see Chapham, 1993)[4] that does not do justice to the complexity of this perhaps most refined institution extant today. Except for the highest varna of the Brahmins, the remaining varnas consist of numerous jatis. But it is useful in the sense that the categories so derived allow for a pan-Indian examination of issues related to caste. It can be seen as the conceptual framework for the actual practice of the society stratified by jati (Sau, 1999).

Although very prominent, and studied both intensively and extensively by anthropologists, predominantly in field studies but also on a theoretical level, the meaning of the caste system for development has not been scrutinised thoroughly by economists. It is predominantly both the alleged stability of this unique institution and, in connection with that, fatalism, the presumed tendency of the poor to ascribe everything to their *karma*, which led many economists to the conclusion that the caste system impedes modernisation of the Indian economy (see, for example, Akerlof, 1976; Kuran, 1987; Lal, 1988; Marx, [1872] 1971; Myrdal, 1968; Olson, 1982; Scoville, 1996; Weber, 1972; and a comparative survey of economic, historical, and anthropological literature in Subrahmanyam, 1996). They blurred both concepts with the resulting lack of an adequate picture of Indian reality as known through anthropological fieldwork. In order to establish through economic theory the rigidity and drawbacks of the caste system, they overlooked the fact that it has always been much more open, flexible, and adaptable.

The actual meaning of karma is action or deed that also influences current and future lives, but it is often (mis-)interpreted simply as fate, which defies the ability of individuals to influence their present life, which is inconsistent with the actual Hindu philosophy as understood, for instance, by Thapar (1990).

Parry (1996), firmly rejecting Weber's thesis of the "spirit of capitalism" being absent in India, states that the ethical preconditions for the emergence of capitalism have been much more favourable in India than they actually were in Europe. This might be seen as implicitly subscribed to by Lal (1988), who argued that the caste system initially was a highly efficient institution and very much in favour of economic development, embodied quite early in a high level equilibrium which then was maintained at stable conditions over millennia by still entrenched hierarchies or distributional coalitions in the terms of Olson (1982). The recognition of commerce, trade, and other sources of accumulating wealth being in conformity with the religious doctrines, which are definitely culturally determining, on whichever element the emphasis is placed, allows for identification of a climate in traditional India, be it in ancient times or in remote areas today, which was unmistakably favourable to generating a capitalist spirit.

There is a widespread misconception of the Indian, or better "Hindu", attitude towards secular affairs, at least if one tries to locate the source of fatalism and "accommodation", to use Galbraith's ([1979] 1993) notion, in the roots of the cultural and religious traits as manifested in the ancient scripts. This inference was initially proposed by Weber[5] and is known also as the karmic view of Indian society. This picture erroneously propagates the pursuit of religious duties and the outright rejection of material wealth as the basic components of "Hindu" belief. Thence, according to this perspective, supposedly originates the ignorance of the, indeed, very religious Indian population of (technical) change and innovation that could otherwise bring progress to and enhance the welfare and well-being of the deprived population. The presumption underlying this image is a traditional society with a well-balanced power structure in which innovations of whatever nature are deemed to be a threat to the existing equilibrium.[6] These scholars ascribed the stagnation in what Lal calls a "high level equilibrium trap" (1998a: 34) mono-causally to the extant ideologies of "Hinduism".

After, I hope, convincingly debunking the perception of (Western) economists of the values of Indian society as allegedly impeding development, I will turn to the investigation of actual behaviour. Such an analysis is much more anthropological than economic.

3.2 Behaviour

The single most important fact to state with reference to the translation of values and beliefs into action and behaviour is that there are usually large deviations. The motivation for such deviations has been analysed with game-theoretic approaches as "preference falsification" (Kuran, 1987).

One reason for the dissonance between ideological beliefs and actual behaviour is to be found in the philosophical conceptualisation of Hinduism. Whereas many scholars are intrigued by the picture of an overriding principle of Hindu religion like karma, often translated as fate, such an idea does not in fact exist. Rather karma has to be complemented, if not substituted, by *dharma*, perhaps best translated as "sacred duty",[7] as the most important principle in determining and understanding the behaviour of Hindus.

Heesterman restates the view of merited sociologists such as Louis Dumont that dharma is the ever-present moral order which guides the people rather than commands them ("[L]e dharma règne de haut sans avoir, ce qui lui serait fatal, à gouverner", cited in Heesterman, 1984: 77; see also Srinivas, 1978). This aspect of dharma can be seen as the abstract, all-encompassing philosophy of Hinduism.

But dharma is not a monolithic concept; it has to be seen as contingent on the person and the group they belong to (Morris, 1967). There is another implication of dharma which is much more practical: it is one's duty obligated to a certain caste (svadharma, i.e. one's own dharma) in order to maintain the cosmic harmony. Parry speaks of the seemingly very similar notion of jati-dharma as "the code of conduct which is an aspect of his nature" (Parry, 1996: 78). Again we have an ambivalent meaning of the proper word; for example Gelblum (1993: 38) interprets it as the "concept of the individual's duty", which is implicitly reciprocal of the concept of rights.[8] It is in any case of relevance to the individual insofar as it conveys the moral or right actions conforming to the caste one belongs to.

It is quite interesting to note the parallels between dharma, jati-dharma and svadharma on the one hand and "conventions, norms of behaviour and self-imposed codes of conduct" on the other hand (North, 1990: 23). In this sense the social system of Hindu civilisation could be plausibly subsumed under a general categorisation with regard to informal institutions.

Upper-caste groups, such as Brahmins and merchant and ruling groups, have traditionally discriminated against lower-caste groups, but the ranking of upper- and lower-caste groups has varied by region and through time.

"If the stability of the caste order could not hinder property differentiation, it could at least block technological change and occupational mobility... [E]ven today, the very fact that new skills and techniques actually lead to the formation of new castes or subcastes strongly handicaps innovation. It sustains tradition no matter how often the all-powerful development of imported capitalism overrides it." (Weber, 1958: 104)

Although there has always been upward (and downward) mobility, one could say that the principal varna providing economic services such as merchants or entrepreneurs was that of the Vaishyas (Rutten, 2002), complemented, of course, by the minority communities of *Parsis*, *Jains* and *Sikhs* (for example Tripathi, 1992).

Brahmins, on the other hand, were traditionally priests and teachers or in related professions. These related professions cover all the tasks necessary to perform the various religious rituals. These consist mainly of studying, reciting, and handing down the sacred texts, but include auxiliary sciences such as grammar and astronomy, as well as mathematics and geometry (in order to optimally construct the altar for sacrifices) (Stietencron, 1995). Moreover, the Hindu-Brahminical education system encompassed medicine, literature, philosophy, and logic (Das, 2000; Gosalia, 2000). Hence, there are many disciplines that are very useful for intellectually challenging professions in the so-called knowledge economy such as sciences or research-related pharmaceuticals, biotechnology, or software. This tradition being handed down from one generation to the next for decades or even centuries would place descendants in a privileged position regarding such professions and, thus, be an example of (economic) culture as summarised in the previous section. Therefore, even if Brahmins have monopolised learning, as some argue, there might be a positive impact on the Indian economy in the "knowledge age" (Das, 2000: 153).

Especially with regard to traditional professions like artisans, this division of labour seems to be still perfectly in place. There is evidence from various field studies, both economic as well as sociological and anthropological, that this holds true.[9] Moreover, such a network of interdependent producers and traders adhering to their customary occupations can be described as a cluster. In the traditional footwear cluster of Agra, a major factor in the successful mastering of crises is the extent of vertical relationships (Knorringa, 1999).

However, with regard to the urban, and especially the metropolitan, areas of India, this traditional aspect of the culture derived from religion is being undermined by various factors, most notably industrialisation and occupational diversification in general (Srinivas, 1978). In particular, caste is being superseded by issues of class (and ethnicity), more so among the upper than the lower castes (Béteille, 1996).

Whatever might be the importance of these moral values today, it is noteworthy how they are supposed to have spread during the past millennia in a process described as "Sanskritisation" of the lower castes; that is, the imitation of customs and rites as followed by the Brahmins, to the extent possible in terms of ability to perform these rites (Srinivas, 1978).

As already said, the most widespread inference made does not take into account the internal dynamics of caste and its adaptability and tolerant attitude towards external institutional changes, be they political or legislative or anything else (Osborne, 2001). Nevertheless, this stance is usually taken by economists who ignore the insights from history and anthropology evidently showing the opposite. They account for neither the upward mobility of previously lower castes through economic success nor the process of Sanskritisation as inherent to caste. Instead the caste system is seen as inseparably interrelated with the Hindu religion, despite the fact that it hosts other religions and sects too, albeit as subdivisions, jatis, being ranked according to the prevailing circumstances like, for instance, economic success.

Closely related to the flexibility of the system with regard to the mobility of castes, or better jatis, almost as a precondition, is the emergence of new occupational activities. The jati-dharma has been almost fixed for ancient castes and jatis through scriptural tradition. But there cannot be prescriptions for all the newly evolving subcastes, just as there cannot be explicit contracts specifying every contingency. Thus, in modern Indian society there is some vagueness concerning the future adherence and obedience to dharma, because there is no (religious) authority legitimated to declare such a social code of conduct. Furthermore, the developments observed reveal a certain tendency to reverse the process Max Weber has described as "the transformation from ethnicity to caste" (Fuller, 1996: 22), which Dumont labelled "substantialisation of caste" (quoted in Béteille, 1996: 172; Fuller, 1996) and anthropologists more broadly have named ethnicisation. It might be particularly meaningful in India as it dissociates class from caste (Béteille, 1996; Fuller, 1996; Platteau, 1995).

Recently, there is some more than anecdotal evidence that the structure of new Indian enterprises is determined by Brahmins rather than Vaishyas, the traditional merchant caste; this evidence is analysed in the next section. It might result from the fact that Brahmins have been involved not only with the profession of priesthood but more generally with activities relating to knowledge and wisdom. In earlier times Brahmins had a much more negative attitude towards business, trade, and commercial castes in general. Lal (1995) called this attitude atavistic and described the Brahmins as primarily protecting their status.

An interesting comparison between values and practices is the GLOBE project, which identifies cultural clusters worldwide. India, as part of the South Asia cluster, is distinguished as highly group oriented, humane, and male dominated. Regarding business strategies, the study finds that South Asian managers focus on a combination of knowledge, action, and devotion (Chhokar, 2002; Gupta et al., 2002).

3.3 South India

With regard to South India there are a few notable, somehow contradictory deviations compared to the rest of India. Firstly, caste is perceived to have been imported by (Brahmin) Indo-Aryan-speaking migrants from the North.[10] Therefore the position of the Brahmins as representatives of this hierarchical order seems to be more exposed in the southern states, particularly Tamil Nadu, but also the neighbouring Karnataka and Andhra Pradesh. However, there is also an indigenous Dravidian culture with its own languages, symbols, and sacred texts. These South Indian cultural elements had to be balanced by the Brahmins against their own Sanskritic traditions. On the other hand, there have always been high-caste non-Brahmins within the indigenous Dravidian population who not only were engaged in the learning of the Dravidian texts but "were adept in Sanskrit learning as well" (Stein, 1999: 52). Hence,

Brahmins are to be seen as mediators who provided for the diffusion of Sanskritic knowledge rather than as monopolists. Thus, the foundations for a knowledge-based society have apparently long existed in South India and, moreover, have been much more diffused throughout the whole society. Moreover, political movements in favour of the backward groups of Indian society started much earlier in the South and led to a more equal pattern there as opposed to the more traditionally dominated northern states (Jaffrelot, 2002).

Secondly, the indigenous population of the South is said to be much more homogeneous and not to display the two middle caste groups of *Ksatriyas* and *Vaishyas* to the same extent as in the North (Dirks, 1996; Stein, 1999). In addition, there was a further distinction between so-called "right hand" and "left hand" divisions in at least three of the four southern states, adding to the complexity of the hierarchy (Srinivas, 2003). Both Brahmins and high-caste non-Brahmins have been excluded from the occupations of these two caste groups, which included different kinds of traders and merchants in both agricultural and non-agricultural occupations. The absence of the warrior castes of the Ksatriyas in particular resulted in a generally more peaceful and contemplative society, one reason often cited for the higher political stability of South Indian states (Stein, 1999).

This considerable emphasis on learning is reflected by the higher proportion of institutions of tertiary education and a greater presence of higher education establishments in the South Indian states (D'Costa, 2003). This is particular revealing when taking into account the lower economic status of these states compared with those of North India (Chalam, 2000). Generally, mathematics and other pure sciences are said to confer a high status on people proficient in them, since they are intellectually the most demanding (Krishna et al., 1998).

What is particularly striking is the large number of technical and engineering colleges in the four southern states (D'Costa, 2003). Although the numbers vary according to the definition of a college used by different sources, the figures remain relatively stable. Dossani (1999), for example, identifies one of the strengths of the South Indian states as their technical capacity, with 400 out of 600 technical colleges located here. Similarly, Arora and Athreye (2002) report that roughly 50 per cent of all Indian engineering colleges and enrolment as well as 79 per cent of privately financed colleges are in South India, as against the national average of 69 per cent (table 9.1).

Arora and Athreye find the latter figures in particular startling and hypothesise that they might be related to cultural and political factors. Chalam (2000) more explicitly links this fact to social and cultural move-

Table 9.1 Number of engineering colleges and enrolment

Region	No. of engineering colleges	National share in engineering colleges	Enrolment (sanctioned capacity)	National share in enrolment	Percent share of national population	Share of self-financed colleges
Central	50	7.54%	9,470	6.05%	–	52%
East	25	3.77%	4,812	3.07%	25.8%	26%
North (incl. North-west)	140	21.12%	25,449	16.26%	31.3%	42%
West	140	21.12%	34,165	21.83%	19.6%	74%
South (incl. South-west)	308	46.46%	82,597	52.78%	23.2%	79%
Total	663	100.00%	156,493	100.00%	100.00%	69%

Source: Arora and Athreye (2002).

ments (see previous section) that tried to break away from the traditional and superstitious customs in order to arrive at more humanitarian values.

Taken together, the southern part of India seems to exhibit a more intensive regional culture not only of learning, quite literally rather than in the sense of the regional development literature, but also of innovation (see chap. 8 in this volume). Apparently, this attitude is a solid foundation for the absorptive capacity necessary to adapt to new technologies (Lateef, 1997).

4 The Indian ICT industry

This section provides an overview of the Indian software industry. It is basically a survey of the recent literature in economics and geography dealing with IT in India. This review could also be described as a summary of the usual analysis of the success of Bangalore, which will be supplemented by a cultural explanation later on. This cultural argument is supported by the findings of previous interview-based studies which are discussed in this section. It should be highlighted that this argument is being *added* to the conventional analysis of the software agglomerations.

4.1 Overview

The Indian ICT industry mainly consists of a broad spectrum of software development enterprises. It contains the most successful branches of the services industry in the Indian economy. The growth, particularly of software exports, has been at an astonishingly high compounded annual rate of 46 per cent for the last decade (1990/1991 to 1999/2000) (Parthasarathi and Joseph, 2002), which is quite respectable with respect to the same industry in other (developing) countries. It is for this reason that other emerging economies have tried to replicate the "Indian model", similarly to the orientation towards Silicon Valley by industrialised countries (Arora et al., 2001b). There is a high demand abroad for Indian support in establishing software technology parks (STPs) on the Indian model, most of the demand coming from countries of the Asia-Pacific region such as Hong Kong, Singapore, Korea, and China. This stems from the fact that, although some scholars describe the involvement of the Government of India as "benign neglect" (Arora et al., 2001b: 25) rather than actively stimulating business, it did in fact recognise the importance of supporting the software sector in general, and exports in particular, as early as 1972 (Parthasarathi and Joseph, 2002) with the initiation of an export processing zone near Bombay (Bajpai and Shastri,

1998). Other supportive policies, such as establishing the prestigious Indian Institutes of Technology, have been of critical value to the evolution of the software industry (ibid.). Subsequently, the first STPs were established in 1990 (Parthasarathi and Joseph, 2002). But there could be more policy initiatives, such as encouraging investment by non-resident Indians. For example, from 1991 to 1998 India received as foreign direct investment only about one fourth of the amount China received in 1998 alone (Klein and Palanivel, 2000).

There are some authors who are rather critical of the innovative capabilities of the Indian software industry (ibid.). Their argument is that the major activities consist of data entry, on-site project work (mostly in the United States), or others jobs rather low in the value chain (Arora et al., 2001b). India could excel with these activities as long as there is a cost advantage. But these activities are being increasingly automated. The relatively unimportant domestic market is usually identified as the major shortcoming (Bajpai and Shastri, 1998).

On the other hand, the quality of the software exporting firms is certified at high levels. Indian firms provided the largest number of ISO 9000 certified companies worldwide in 1998 (ibid.), and more than half of those certified in 2001 (Arora et al., 2001a: 1283). Moreover, worldwide, India has the largest number of enterprises certified to Level 5 of the Carnegie Mellon University's Software Engineering Institute (SEI) Capability Maturity Model (CMM) (Bajpai and Shastri, 1998); recently Wipro Technologies has become the first company to obtain Level 5 both of the People Capability Maturity Model and of the CMM-SEI (Yue et al., 2001).

Even if the export performance is overstated, as some argue (Parthasarathi and Joseph, 2002), and individually disappointing results lead to falling share prices, this is in sharp contrast with the rest of the economy. Since independence more than 50 years ago, the Indian economy has grown at only 3.5 per cent, the so-called "Hindu Rate of Growth" that implies "deep cultural factors" (Bhagwati, 2001).

4.2 Results of interview data

Certainly, there is no doubt about the purely economic factors that have contributed to the successful evolution of the Indian software industry. It seems quite obvious, for example, that the liberalisation initiated in the 1980s and accelerated to a certain extent in the first half of the 1990s has made its contribution. However, regarding the software industry, which reached a critical size only in the 1990s, one finds that its locations are relatively unevenly distributed. Basically the industry is clustered in the three South Indian regions of Bangalore, Hyderabad, and Chennai (formerly Madras), as well as the western Indian cities of Mumbai (formerly Bombay), Pune and Ahmedabad, and around the capital New Delhi.

Thus the question that is addressed here is why some regions, largely in the southern (and western) parts of India, are more successful than their counterparts in the rest of the country. The hypothesis is that, beyond economic and geographical aspects, cultural influences come into play and have a significant impact upon the economy. These cultural influences are approximated through the variables of the geographical and social origin of the persons interviewed.

The approach is a very simple, qualitative one looking at the interview data of previous studies of the Indian software industry. The studies analysed are those of Tschang (2001), Saxenian (1999), Bajpai and Shastri (1998), and Arora et al. (2001b). No econometrics are employed, though this should be done in future research. Nor were correlations tested, for the number and clarity of the data were not entirely satisfactory and only frequencies based on the variables of location and social and regional background or origin were calculated. What follows are the first preliminary results.

All interviewees were key entrepreneurs, managers, or administrative staff. The information provided in the appendices to these studies is not uniform, thus the sample sizes are not the same for all distributions. However, altogether there are more than 200 entries, nearly all of which mention the name of the interviewee, which is crucial to my findings.

Almost all studies claim to cover the entire software industry and do not specify a particular regional focus. However, analysing the interview data of these studies one finds a bias towards South Indian (and, to a lesser extent, western Indian) locations as the major centres of this industry. More than 90 per cent of the interview partners came from firms or authorities in southern or western India (fig. 9.1). Of those, more than 50 per cent were from Bangalore and its surrounding state Karnataka (fig. 9.2). This finding is supported by Heeks (1998), but is in contrast with some of the studies asserting that Bangalore is not the centre of the software industry, but is rather losing its former status as "the Silicon Valley of India" (Arora et al., 2001a: 1272). Presumably, the industry is still at an early stage in which the distribution of companies varies highly. Therefore the merit of being the number one location might change between Bangalore, Bombay, and probably Hyderabad. Of course, there are the usual explanations of university–industry linkages with the Indian Institutes of Technology (IITs) and the establishment of software technology parks close to the IITs, as well as historical circumstances which led to the initial localisations.

But, if one takes a closer look at the names of the interviewees, there is another remarkable finding. As shown in the previous section, customary

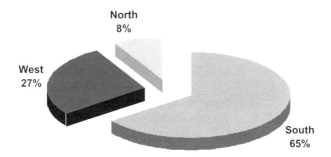

Figure 9.1 Regional distribution of interviewees in India.
Source: Based on Bajpai and Shastri (1998), Saxenian (1999), Tschang (2001), and Arora et al. (2001b).

values and traditions are still adhered to in India, especially in South India. This phenomenon finds expression in the fact that one can tell the ethnic and social, but not necessarily the economic, background of many Indians from their name (Deshpande, 2002). This is done by the use of anthropological literature and an Oxford dictionary with a supplement of Indian terms. Although the results seem to be very clear, I must state that one has to be very careful interpreting these results.

The findings for the interviews show that almost 70 per cent of the interviewees are Brahmins, irrespective of the location within India (fig. 9.3). The ethnic background is slightly more ambiguous, with roughly 50 per cent of the people interviewed being from South India and an additional quarter from the otherwise underrepresented Hindi heartland in the Centre/North (fig. 9.4). However, there is again an overrepresentation of the South compared with its share in the national population (table 9.2).

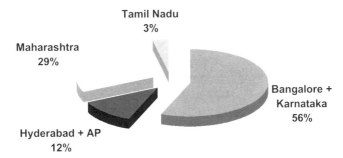

Figure 9.2 State-wise distribution of interviewees in South and West India
Source: Based on Bajpai and Shastri (1998), Saxenian (1999), Tschang (2001), and Arora et al. (2001b).

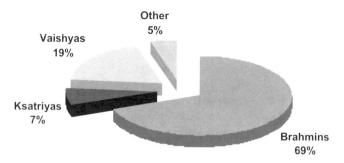

Figure 9.3 Social background of interviewees in India
Source: Based on Bajpai and Shastri (1998), Saxenian (1999), Tschang (2001), and Arora et al. (2001b).

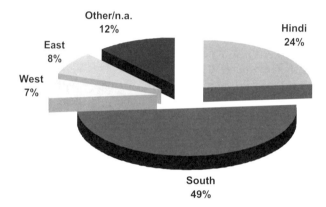

Figure 9.4 Ethnicity of interview partners in India
Source: Based on Bajpai and Shastri (1998), Saxenian (1999), Tschang (2001), and Arora et al. (2001b).

Table 9.2 Ethnicity of interviewees in the Indian software industry and their share in the national population

Region of birth	Proportion in the software industry[a]	Percent share of national population[b]	Percentage of over-/under-representation
South India	49	23.2	+111%
West India	7	19.6	−64%
North India	24	31.3	−23%
East India	8	25.8	−69%
Other	12	–	–

a. Own calculations, based on Tschang (2001), Saxenian (1999), Bajpai and Shastri (1998), and Arora et al. (2001b).
b. From Dossani (2002).

To summarise, the findings are that the majority of the key people in the Indian software industry are located in South India, are Brahmins, and come from a South Indian background in terms of ethnicity or family affiliation. These findings are supported by the recent study by Saxenian, Motoyama and Quan (2002), which highlights the importance of South India as a destination for deliberately planned investments and returns to their home country by Silicon Valley Indians.

Before I venture into a discussion, I should state once again that the results should be interpreted with caution because the data are not complete and sometimes not entirely consistent. However, the strength of the result allows for some provisional interpretations that call for further research in this direction.

The regional distribution seems to be influenced by historical and geographical factors, at least to a certain extent. Similarly, ethnic background might simply be due to the same reasons. The historical factors are the early localisation of science and technology related research and teaching institutions in Bangalore as an ideal place in terms of climate and infrastructure to conduct scientific research in sensitive areas such as defence and electronics. To qualify what has been said earlier, the southern centres are not the only centres, but they seem to be the dominant ones. That is, there also are STPs in North India together with successful local firms and regional offices of South Indian companies.

What is more surprising, however, is the distribution of socio-cultural and ethnic descent. There have never in Indian history been so many entrepreneurial and managerial Brahmins as are now seen in the software industry. Generally speaking, Brahmins were rather associated with priestly tasks, government jobs, administration, and landholding. On the other hand, Brahmins as members of the priestly caste were always connected to various activities related to knowledge, learning, and teaching. That might explain why they seem to be well suited to knowledge-intensive industries like that of software (Das, 2000; Zingel, 2000). Moreover, the combination of subjects emphasised by a Brahminical syllabus seems to be especially apt for software programming, which requires not only mathematics but also language or grammar. In addition, even though few Indian software companies produce software products or packages, codification is still important.[11] Most of the firms are engaged in outsourcing from Western or multinational companies. In order to communicate effectively between the two corporations involved, a common language platform is essential. Therefore, at least some work must be done in a codified manner (Grimaldi and Torrisi, 2001). Here, not only a particular Brahminical tradition of analytical and methodical thinking comes into play, but also the way Indian software experts are trained, that is according to the International Standards Organization methodology (Nicholson and Sahay, 2001), implying a high degree of codified knowledge. Vice versa, the software industry is particularly suitable for Brahmins, for it is an industry where they do not run the risk of pollution, which has traditionally been a major concern of many Brahmins (Zingel, 2000).

Also unexpected is the relative underrepresentation of Vaishyas. As stated above, they have always been the entrepreneurial castes of the Hindu population (Dorin, 2003). Although there have always been other entrepreneurs (Tripathi, 1992), not even in combination with the Parsis and Jains do they make up their usual share. While it seems plausible that there is a high percentage of Brahmins in the industry workforce, the industry's leadership also seems to be dominated by South Indian Brahmins (Fromhold-Eisebith, 2000; Merchant, 2002). This is fascinating insofar as South Indians as a social group used to be excluded from the entrepreneurial pool of the Indian business houses (Kapur and Ramamurti, 2001).

This also leaves unexplained why so few northern Indians act as entrepreneurs in the relatively young software industry. It could be assumed that they have been lured to the southern regions by high profits. Another explanation lies in the attitude of the traditional merchants and trader class towards risk and quick profits. They often prefer the latter and avoid taking risks, thus forgoing higher profits in the longer term (Frederking, 2002). But probably there are entrepreneurs in the North who were just missed by the interviewers.

Apart from the association with knowledge and learning, there is another, socio-cultural factor that might explain the dominance of Brahmins in the South. As indicated in the previous section, this factor is ambiguous. Not being an indigenously Aryan-speaking people, South Indians were fiercely opposed to the Brahmin domination of the North. It has been argued that the non-Brahmin society was relatively homogeneous, which might partly explain why in South India the two middle varnas were absent. This absence in turn could have led to an even more dominant position of the Brahmins in the southern parts of India, which dominance was then compounded by land ownership and political power (Dirks, 1996). The contradiction regarding industries such as software lies with South Indian society apparently being more egalitarian in terms of traditionally opening knowledge to broader sections of the population, as witnessed by a higher proliferation of education institutions in modern times. But this could eventually have become dominated by the political power of the dominant castes. Deshpande (2000) calls a cumulative advantage the fact that the upper castes today are in such a strong position that, in order to retain their privilege, they no longer need the customary inheritance of status (see also Nafziger, 1975). Such a dominant position in administration could have been used in order to ensure a more than proportionate share of Brahmins in high schools and universities. It would be worthwhile undertaking some more research analysing the networks of the IT industry that have been built through the educational system.

5 Conclusion

This chapter has looked at the success of Indians in the global IT industry from a new perspective. The view taken here suggests that, beyond the economic and geographical factors that have been extensively discussed elsewhere, cultural aspects play an important role. An attempt has been made to operationalise culture through a pair of variables. These variables are caste, in the sense of varna as an aggregation of occupationally inherited jatis, and ethnicity. In spite of the shortcomings such an aggregation inherently possesses, there is some evidence that within India the traditionally assigned occupations are losing their importance. Nevertheless, caste continues to play its part in the society, primarily in politics. To a certain extent, the second variable, ethnicity, is correlated with the first, although through aggregation much of this connection is lost. This stems from the fact that the regional variations within the system of jatis are not reflected through varna. However, interpreting the two varnas most featured in the interview data, the Brahmins broadly as priestly and knowledgeable and the Vaishyas as those occupied with commerce and business, displays a socially powerful result. The once dominant group of entrepreneurs has not retained its customary share in business in the evolving software industry, which is now largely controlled by Brahmins. This statement needs to be qualified for the small sample size and some incompleteness of data in a few cases. Also, there might be an ex-post selection bias stemming from the concentration on the most successful firms in the industry in the South, overlooking social and cultural changes taking place in other regions. However, the results indicate a worthwhile direction for further research.

Summarising, it can be said the rise of the software industry in South India stems from the combination of a broad human capital base and the geo-physical circumstances, at least in the case of Bangalore, stimulating government policies, which to a certain degree have been unintentional, and the involvement of non-resident Indians. But most of all, the rise of the industry appears to be the result of the generally more hospitable attitude towards education and learning in South India and the eventual development of an infrastructure in (higher) education. Also the software industry seems to be clearly dominated by South Indian Brahmins, which, in turn, may explain the spatial concentration of this industry. Moreover, the geographical origin of migrants seems to play another important role, since most of the flows through the transnational networks are directed towards South India, which might become a positive feedback process thereby increasing the regional concentration of the Indian software industry. This process seems to be already well under way, providing in particular Bangalore, and to a lesser extent Hyderabad, with the competitive advantage for research-intensive industries in a knowledge economy.

As already mentioned, this chapter is the first rather theoretical part of a larger project analysing the effect of socio-cultural influences on the evolution of the Indian software industry. Further research will be undertaken to assess the impact of social networks among the South Indian communities in the software industry both within India and abroad. We have already interviewed managers of Indian software firms in Frankfurt and found only limited evidence of networks and collaboration, probably due to the primarily marketing functions of these offices. It is planned to conduct semi-structured interviews with company executives as well as university faculty members in Bangalore at the end of 2003. The result is expected to indicate a process of social networking starting at university or college level, thereby supporting the cultural influence argument. These networks are thought to encourage entrepreneurship by both faculty members and the employees of larger, often multinational, companies through mutual support and cooperation. Hence innovation in the IT industry would occur to a large extent at the research level. If this holds true it would not be surprising that the Indian software industry is concentrated in South India, since the number of education institutions is much higher there. However, this would implicitly stress that software is primarily a people-centred business and, again, place relatively little emphasis on other "hard" locational factors.

Notes

This is a revised version of a presentation delivered at the United Nations University–Institute for New Technologies, Maastricht, the Netherlands. Earlier versions of this paper have been presented at conferences at New School University, New York, and the Center for International Development at Harvard University, Cambridge (Mass.), and the European Summer School on Industrial Dynamics at Carghese. I would like to thank all the participants in these conferences, in particular Peter Knorringa, Sunil Mani, Lynn Mytelka, Rajah Rasiah, and Henny Romijn; Alberto Araoz, B. Bowonder, Uma Chandru, and Smita Srinivas; and Paul David and Dominique Foray, for helpful comments. All remaining errors are, of course, solely my responsibility.

1. Gupta mentions a taxonomy collected as early as 1952 which gives 164 definitions of culture; presumably there is a multiple of this number today (Gupta, 1994: 2).
2. Grondona has a typology of 21 factors.
3. Dumont (1980) regards the fact that nobody is excluded from the societal structure, contrary to (medieval) Western class conceptions, as the most outstanding feature of the caste system.
4. M. N. Srinivas (1978) mentions an earlier "calculation" by Ghurye "that there are 2000 subcastes (jatis) in every linguistic area". Chapham (1993) even speaks of 3,000 subcastes for the major groups, and of 90,000 endogamous marriage groups.
5. "Ever since Max Weber's analysis of Indian society, many Western (and Indian) social scientists have interpreted social institutions such as caste and the extended family as oppressive, in the sense of hindering the growth of such personality traits as 'independence', 'initiative', 'persistence' and 'achievement motivation' in the individual" (Kakar, 1981: 10).
6. Warner (1970) defends Weber's ideas against critics such as Morris (1967) who argue in favour of a secularly oriented mentality that is also responsive to many forms of incentives as offered, for instance, by the market mechanism. The latter view is based partly on studies of industrial entrepreneurs in western India. See the discussion in Subrahmanyam (1996) and the chapter by Singh (1989) on "Relevance of Max Weber for the Understanding of Indian reality". The conclusion is that, in spite of many interpretations that have mistaken Indian reality, in terms of sociological methodology, Weber's approach can still be regarded as extremely useful. Accordingly, Weber's idea of "interpretative understanding" as part of his methodology is especially valuable for Indian sociologists, despite his misplaced emphasis on the meaningfulness of cultural factors.
7. Thanks are due to Prof. Burkhard Schnepel, now of the Institute for Social Anthropology, Martin Luther-University, Halle, for this remark commenting on an observation made during several field studies in India.
8. This aspect was emphasised by Mahatma Gandhi, who considered varnadharma as the most important and attached to it the importance of the group as opposed to the individual and the primacy of duties over rights (see Béteille, 1996: 160).
9. See Nafziger (1977, 1986, 1994) for economic studies conducted in the southern state of Andhra Pradesh; Rutten (1999, 2000) for a sociological perspective on the western state of Gujarat; and Reiniche (1996) for an anthropological field study of a merchant caste in South Indian Tamil Nadu.
10. Stein (1999) points out that there is no evidence for such an invasion or migration of substantial dimensions.
11. I would like to thank Sunil Mani and Rajah Rasiah for clarifying my understanding of the importance of codification in the software industry.

REFERENCES

Abramovitz, M. (1986) "Catching-up, Forging Ahead, and Falling Behind", *Journal of Economic History* 46(2): 385–406.

Akerlof, G. A. (1976) "The Economics of Caste, of the Rat Race and Other Woeful Tales", *Quarterly Journal of Economics* 90(4): 599–617.

Arora, A. and S. Athreye (2002) "The Software Industry and India's Economic Development", *Information Economics and Policy* 14(2): 253–273.

Arora, A., V. S. Arunachalam, J. Asundi and R. Fernandes (2001a) "The Indian Software Services Industry", *Research Policy* 30(8): 1267–1287.

────── (2001b) "The Globalization of Software: The Case of the Indian Software Industry", report submitted to the Sloan Foundation, Pittsburgh: Carnegie Mellon University.

Bajpai, N. and V. Shastri (1998) "Software Industry in India: A Case Study", Development Discussion Paper No. 667, Harvard: Harvard Institute for International Development, University of Harvard.

Bayly, S. (1999) *Caste, Society and Politics in India from the Eighteenth Century to the Modern Age*, Cambridge: Cambridge University Press.

Béteille, A. (1996) "Caste in Contemporary India", in C. J. Fuller, ed., *Caste Today*, Delhi: Oxford University Press, pp. 150–179.

Bhagwati, J. (2001) "Growth, Poverty and Reforms", *Economic and Political Weekly* 36(10): 843–846.

Chalam, K. S. (2000) "Human Resources Development in South India", *Journal of Social and Economic Development* II(2): 291–313.

Chapham, G. (1993) "Religious vs. Regional Determinism: India, Pakistan and Bangladesh as Inheritors of Empire", in D. Arnold and P. Robb, eds., *Institutions and Ideologies*, Richmond: Curzon Press, pp. 5–34.

Chhokar, J. (2002) "Leadership and Culture in India: The GLOBE Research Project", in R. House and J. Chhokar, eds., *Studies of Managerial Cultures in 12 Countries*, Calgary: Haskayne School of Business, University of Calgary, http://www.ucalgary.ca/mg/GLOBE/Public/Links/india.pdf.

Das, G. (2000) *India Unbound – The Social and Economic Revolution from Independence to the Global Information Age*, New York: Anchor.

D'Costa, A. (2003) "Uneven and Combined Development: Understanding India's Software Exports", *World Development* 31(1): 211–226.

Dehejia, R. and V. Dehejia (1993) "Religion and Economic Activity in India: An Historical Perspective", *American Journal of Economics and Sociology* 52(2): 145–153.

Deshpande, A. (2000) "Does Caste Still Define Disparity: A Look at Inequality in Kerala, India", *American Economic Review, Papers and Proceedings* 90(2): 322–325.

────── (2002) "Assets versus Autonomy? The Changing Face of the Gender–Caste Overlap in India", *Feminist Economics* 8(2): 19–35.

DiMaggio, P. (1993) "Culture and Economy", in N. Smelser and R. Swedberg, eds., *The Handbook of Economic Sociology*, Princeton, N.J.: Princeton University Press, pp. 27–57.

Dirks, N. (1996) "Recasting Tamil Society: The Politics of Caste and Race in Contemporary Southern India", in C. J. Fuller, ed., *Caste Today*, Delhi: Oxford University Press, pp. 263–295.

Dorin, B., ed. (2003) *The Indian Entrepreneur: A Sociological Profile of Businessmen and Their Practices*, New Delhi: Manohar.

Dossani, R. (1999) "Accessing Venture Capital in India", Conference Report, Asia/Pacific Research Center, Stanford, Calif.

────── (2002) "Chinese and Indian Engineers and Their Networks in Silicon Valley", Asia/Pacific Research Center, Stanford, Calif.

Dumont, L. (1980) *Homo Hierarchicus: The Caste System and Its Implications*, Chicago and London: The University of Chicago Press.
Frederking, L. (2002) "Is There an Endogenous Relationship between Culture and Economic Development?", *Journal of Economic Behavior and Organization* 48(2): 105–126.
Fromhold-Eisebith, M. (2000) "Eine asiatische Technologieregion als neue Variante des kreativen Milieus? Anwendung des Konzepts auf Latecomer-Regionen am Beispiel des indischen Bangalore", *Geographische Zeitschrift* 88(3/4): 147–160.
Fukuyama, F. (2000) "Social Capital", in L. Harrison and S. Huntington, eds., *Culture Matters: How Values Shape Human Progress*, New York: Basic Books, pp. 98–111.
Fuller, C. J. (1996) "Introduction: Caste Today", in C. J. Fuller, ed., *Caste Today*, Delhi: Oxford University Press, pp. 1–31.
Galbraith, J. K. (1993) [originally 1979] *The Nature of Mass Poverty*, London: Penguin.
Gelblum, T. (1993) "Classical Hindu Scriptures", in D. Arnold and P. Robb, eds., *Institutions and Ideologies*, Richmond: Curzon Press, pp. 35–44.
Gertler, M. (1997) "The Invention of Regional Culture", in R. Lee and J. Wills, eds., *Geographies of Economies*, London: Arnold, pp. 47–58.
Gosalia, S. (2000) "Globalisierung und Braindrain im Kontext der Bildungspolitik in Indien", in W. Draguhn, ed., *Indien 2000 – Politik, Wirtschaft, Gesellschaft*, Hamburg: Institut für Asienkunde, pp. 177–194.
Grimaldi, R. and S. Torrisi (2001) "Codified-Tacit and General-Specific Knowledge in the Division of Labour among Firms – A Study of the Software Industry", *Research Policy* 30(9): 1425–1442.
Grondona, M. (2000) "A Cultural Typology of Economic Development", in L. Harrison and S. Huntington, eds., *Culture Matters: How Values Shape Human Progress*, New York: Basic Books, pp. 44–55.
Gupta, A. (1994) *Indian Entrepreneurial Culture: Its Many Paradoxes*, London: Wishwa Prakashan.
Gupta, V., G. Surie, M. Javidan and J. Chokar (2002) "Southern Asia Cluster: Where the Old Meets the New?", *Journal of World Business* 37: 16–27.
Harrison, L. and S. Huntington, eds. (2000) *Culture Matters: How Values Shape Human Progress*, New York: Basic Books.
Heeks, R. (1998) "The Uneven Profile of Indian Software Exports", Development Informatics Working Paper No. 3, Manchester: Institute for Development Policy and Management, University of Manchester.
Heesterman, J. (1984) "Kaste und Karma: Max Weber's Analyse der indischen Sozialstruktur", in W. Schluchter, ed., *Max Weber's Studie über Hinduismus und Buddhismus: Interpretation und Kritik*, Frankfurt: Suhrkamp, pp. 72–86.
Jaffrelot, C. (2002) "The Subordinate Caste Revolution", in A. Ayres and P. Oldenburg, eds., *India Briefing: Quickening the Pace of Change*, Armonk, NY: M. E. Sharpe, pp. 121–158.
Kakar, S. (1981) *The Inner World: A Psycho-Analytic Study of Childhood and Society in India*, Delhi: Oxford University Press.

Kapur, D. and R. Ramamurti (2001) "India's Emerging Competitive Advantage in Services", *Academy of Management Executive* 15(2): 20–32.

Khanka, S. S. (2001) *Entrepreneurial Development*, New Delhi: Chand and Co.

Klein, L. and T. Palanivel (2000) "Economic Reforms and Growth Prospects in India", paper prepared for the Festschrift in honour of Dr. C. Rangarajan, former Governor of the Reserve Bank of India.

Klump, R. (1996) "Einleitung", in R. Klump, ed., *Wirtschaftskultur, Wirtschaftsstil und Wirtschaftsordnung: Methoden und Ergebnisse der Wirtschaftskulturforschung*, Marburg: Metropolis, pp. 9–20.

Knorringa, P. (1999) "Agra: An Old Cluster Facing the New Competition", *World Development* 27(9): 1587–1604.

Krishna, S., A. Ojha and M. Barrett (1998) "Competitive Advantage in the Software Industry: An Analysis of the Indian Experience", paper presented at the IFIP Conference, Bangkok.

Kuran, T. (1987) "Preference Falsification, Policy Continuity and Collective Conservatism", *Economic Journal* 97(September): 642–665.

Lal, D. (1988) *The Hindu Equilibrium: Cultural Stability and Economic Stagnation*, New York: Oxford University Press.

—— (1995) "India and China: Contrasts in Economic Liberalization", *World Development* 23(9): 1475–1494.

—— (1998a) *Unintended Consequences: The Impact of Factor Endowments, Culture, and Politics on Long-Run Economic Performance*, Cambridge, Mass.: MIT Press.

—— (1998b) "Culture, Democracy and Development", Working Paper No. 783, April, Los Angeles: Department of Economics, University of California.

Lateef, A. (1997) "Linking up with the Global Economy: A Case Study of the Bangalore Software Industry", Discussion Paper 96/1997, Geneva: International Institute for Labour Studies.

Lundvall, B., ed. (1992) *National Systems of Innovation: Towards a Theory of Innovation and Interactive Learning*, London: Pinter.

Marx, K. (1971) [originally 1872] *Das Kapital: Kritik der politischen Ökonomie, Band I: Der Produktionsprozeß des Kapitals*, Frankfurt: Ullstein.

Merchant, K. (2002) "Quiet Man with a Powerful Voice", *Financial Times* (London), September 23.

Morris, M. (1967) "Values as an Obstacle to Economic Growth in South Asia: An Historical Survey", *Journal of Economic History* XXVII(4): 588–607.

Myrdal, G. (1968) *Asian Drama – An Inquiry into the Poverty of Nations*, Vol. 3, New York: Pantheon.

Nafziger, W. (1975) "Class, Caste and Community of South Indian Industrialists", *Journal of Development Studies* 11(2): 131–148.

—— (1977) "Entrepreneurship, Social Mobility, and Income Redistribution in South India", *American Economic Review, Papers and Proceedings* 67(1): 76–80.

—— (1986) *Entrepreneurship, Equity, and Economic Development*, Greenwich, Conn. and London: JAI Press.

—— (1994) *Poverty and Wealth: Comparing Afro-Asian Development*, Greenwich, Conn. and London: JAI Press.

Nelson, R., ed. (1993) *National Systems of Innovation: A Comparative Study*, Oxford: Oxford University Press.
Nicholson, B. and S. Sahay (2001) "Some Political and Cultural Issues in the Globalisation of Software Development: Case Experience from Britain and India", *Information and Organization* 11: 25–43.
North, D. (1990) *Institutions, Institutional Change and Economic Performance*, Cambridge: Cambridge University Press.
Olson, M. (1982) *The Rise and the Decline of Nations: Economic Growth, Stagflation and Social Rigidities*, New Haven and London: Yale University Press.
Osborne, E. (2001) "Culture, Development, and Government: Reservations in India", *Economic Development and Cultural Change* 49(3): 659–685.
Parry, J. (1996) "On the Moral Perils of Exchange", in J. Parry and M. Bloch, eds., *Money and the Morality of Exchange*, Cambridge: Cambridge University Press, pp. 64–93.
Parthasarathi, A. and K. J. Joseph (2002) "Limits to Innovation with Strong Export Orientation: The Case of India's Information and Communication Technologies Sector", *Science, Technology and Society* 7(1): 13–49.
Platteau, J. (1995) "An Indian Model of Aristocratic Patronage", *Oxford Economic Papers* 47(4): 636–662.
Reiniche, M. L. (1996) "The Urban Dynamics of Caste: A Case Study from Tamilnadu", in C. J. Fuller, ed., *Caste Today*, Delhi: Oxford University Press, pp. 124–149.
Rothermund, D. (1995) "Epochen der Indischen Geschichte", in D. Rothermund, ed., *Indien: Kultur, Geschichte, Politik, Wirtschaft, Umwelt*, München: C. H. Beck, pp. 77–100.
Ruttan, V. (1988) "Cultural Endowments and Economic Development: What Can We Learn from Anthropology?", *Economic Development and Cultural Change* 36(3): 247–271.
Rutten, M. (1999) "Rural Capitalists in India, Indonesia, and Malaysia: Three Cases, Two Debates, One Analysis?", *Sojourn* 14(1): 57–97.
—— (2000) "Commercialism and Productive Forms of Business Behavior: Rural Entrepreneurs in India, Malaysia, and Indonesia", *Journal of Asian Business* 16(3): 1–25.
—— (2002) "A Historical and Comparative View on the Study of Indian Entrepreneurship", *Economic Sociology, European Electronic Newsletter* 3(2): 3–16.
Sarkar, S. (2000) "Kautilyan Economics: An Analysis and Interpretation", *Indian Economic Journal* 47(4): 62–67.
Sau, R. (1999) "The Hindu Opprobrium, Not Equilibrium", *Economic and Political Weekly* 34 (March 20–26).
Saxenian, A. (1994) *Regional Advantage: Culture and Competition in Silicon Valley and Route 128*, Cambridge, Mass.: Harvard University Press.
—— (1999) *Silicon Valley's New Immigrant Entrepreneurs*, San Francisco: Public Policy Institute of California.
Saxenian, A., Y. Motoyama and X. Quan (2002) "Local and Global Networks of Immigrant Professionals in Silicon Valley", report, San Francisco: Public Policy Institute of California.

Scoville, J. (1996) "Labor Market Underpinnings of a Caste Economy: Foiling the Coase Theorem", *American Journal of Economics and Sociology* 55(4): 385–394.

Sen, A. (1999) *Development as Freedom*, New York: Anchor.

——— (2001) "Culture and Development", paper delivered to the World Bank Tokyo Meetings, 13 December 2000, Tokyo.

Singh, Y. (1989) "Relevance of Max Weber for the Understanding of Indian Reality", in K. Bharadwaj and S. Kaviraj, eds., *Perspectives on Capitalism*, New Delhi: Sage, pp. 211–230.

Srinivas, M. N. (1978) "The Caste System in India", in A. Béteille, ed., *Social Inequality*, Harmondsworth: Penguin, pp. 265–272.

——— (2003) "An Obituary on Caste as a System", *Economic and Political Weekly*, February 1.

Stein, B. (1999) *Peasant State and Society in Medieval South India*, Delhi: Oxford University Press.

Stietencron, H.v. (1995) "Die Erscheinungsformen des Hinduismus", in D. Rothermund, ed., *Indien: Kultur, Geschichte, Politik, Wirtschaft, Umwelt*, München: C. H. Beck, pp. 143–166.

Subrahmanyam, S. (1996) "Institutions, Agency and Economic Change in South Asia: A Survey and Some Suggestions", in B. Stein and S. Subrahmanyam, eds., *Institutions and Economic Change in South Asia*, Delhi: Oxford University Press, pp. 14–47.

Thapar, R. (1990) *A History of India – Volume One*, London: Penguin.

Tripathi, D. (1992) "Indian Business Houses and Entrepreneurship: A Note on Research Trends", *The Journal of Entrepreneurship* 1(1): 75–97.

Tschang, T. (2001) "The Basic Characteristics of Skills and Organizational Capabilities in the Indian Software Industry", ADB Institute Working Paper Series No. 13, Manila: Asian Development Bank.

Warner, S. (1970) "The Role of Religious Ideas and the Use of Models in Max Weber's Comparative Studies of Non-Capitalist Societies", *Journal of Economic History* XXX(1): 74–99.

Weber, M. (1958) *The Religion of India: The Sociology of Hinduism and Buddhism*, Glencoe, Ill.: The Free Press.

——— (1972) [originally 1920] *Gesammelte Aufsätze zur Religionssoziologie, Bd. 2: Hinduismus und Buddhismus*, Tübingen: J. C. B. Mohr [Paul Siebeck].

Yue, C., N. Freeman, R. Venkatesan and S. V. Malvea (2001) *Growth and Development of the IT Industry in Bangalore and Singapore – A Comparative Study*, New Delhi: Sterling.

Zingel, W. (2000) "Indien: Erfolgreich als Dienstleistungsexporteur", in W. Draguhn, ed., *Indien 2000 – Politik, Wirtschaft, Gesellschaft*, Hamburg: Institut für Asienkunde, pp. 343–363.

Index

Note: *Italicised* page numbers indicate figures

advanced materials 36
Advanced Materials Research Centre (Singapore) 68
Aeromot (Brazil) 87
aerospace products 19, 88, 183
　exports 19, 27, 32
　organisations 5
Aerospace Technical Centre (CTA), Brazil 5, 72, 87
Agency for Science, Technology and Research (Singapore) 4
agglomeration
　advantages 139–140, *141*, 145
　and collaboration/networking 145
　and information flows 180–181
　mechanisms 8
　and technological capability/dynamism 144, 147
　see also clusters/clustering; regional agglomeration
Albaladejo, M. 31
Albu, M. 82
Aldaz, E. 31
American Air Force 86
Arbix, G. 127
Argentine Republic 125

armaments 19
Arora, A. 179, 186, 195, 212
ASEAN (Association of South East Asian Nations) 58
Asia 87
　financial crisis (1997) 52
　values debate 203, 205
Asia-Pacific region 70, 183
　clustering 136, 214
　economies 51
Association of South East Asian Nations (ASEAN) 51, 56, 58
Athreye, S. 212
Australia 21, 36
automotive industry *see* international automotive industry

Bajpai, N. 216
Balasubramanyam, V. N. and A. 184
Banco Nacional de Desenvolvimento Economico, Brazil (BNDE) 116, 117
Bangalore software cluster 9–10, 185–186, 221
　development process/techniques 187–188, 192, 196–197
　diffusion of skills and knowledge 190–191

229

Bangalore software cluster (cont.)
  Electronics City creation 191–192
  evolution 183–184
  finance and investment 192–195, 197
  infrastructure 191–192, 197
  innovative capability 179, 189, 190, 191, 197
  labour issues 181–182, 186–191, 197, 198
  markets (domestic/export) 186, 197–198
  and MNCs/TNCs 179, 184, 187, 194, 196, 197–198, 222
  networks 186–189, 222
  offshore projects 192
  quality control 188
  services provision 191–196
  "Silicon Valley" status 179, 216
  social division of labour 186–189
  state policy intervention 196–198
  technology/marketing support services 195–196, 197
  university – industry linkages 195–196
  see also Indian ICT industry; software technology parks (STPs)
BASI (Brazilian civil aircraft manufacturing system of innovation) 5, 6, 78–79, 85
  early features and changes (1969–2002) 93–100
  jetliner production (ERJ 145/ERJ 170) 89, 99–100
  knowledge/production systems 95–100, 102
  post-privatisation (1995–2002) 95–98
  risk-sharing/collaborative agreements 94, 97–98
  suppliers 97, 99
  system changes (1990s) 98–100, 101–102
  technological collaboration 94, 98
BEFIEX (Fiscal Incentives for Exports in Brazil) 121, 124
Bell, R. M. 82, 139
Bennel, P. 167
Bernardes, R. 80, 89, 91
BNDE (Banco Nacional de Desenvolvimento Economico (Brazil)) 116, 117
Bombardier Aerospace (Brazil) 90
Bombay (Mumbai) 183, 216
  export processing zone 214–215
Brazil
  decentralisation 126
  economic/political factors 89, 111–113, 115–119, 120–123
  electronics industry 179
  exports/incentives 16, 121, 125
  foreign exchange and investment 116, 122
  and globalisation 6, 123
  high tech production 102
  industrial policies 77, 80, 81, 130
  industrialisation 78, 115–116, 121–122
  infrastructure 127
  innovation systems 79, 80–81
  institutions 77, 79, 80, 117, 120, 130
  market liberalisation policies 79, 81, 90, 95
  and Mercosur 125
  production and knowledge systems 81
  R&D activity 77, 81, 108, 127
  Real Plan 125
  S&T policies/infrastructure 78–79, 81, 84, 90, 100, 101–102
  technological capabilities/policies 79, 81, 108, 124, 126
  see also BASI; Brazilian automotive industry; Brazilian civil aircraft manufacturing
Brazilian Aerospace Technical Centre (CTA) 78, 87, 94, 100, 101
Brazilian Air Force 87
Brazilian automotive industry 5–6, 108, 115–128
  assembly sector 120, 123
  contribution to "economic miracle" (1968–1974) 121–123, 129
  development strategies 127–128
  development/state intervention period (1952–1974) 115–123, 128–130
  export strategies 122–123, 126–127
  and foreign investment 118, 119, 125–126, 127–128, 129
  generation (1952–1961) 115–119, 128–130
  institution building 120, 130
  "lost decade" (1980–1990) 123–124
  New Automotive Regime (Novo Regime Automotivo) 125, 126, 130
  oil crisis/recession 120, 123–124
  overcapacity problems 119, 120
  parts manufacturers 117–118, 120
  perspectives on growth (1990–2000) 124–126
  politico-economic relations 117–119, 120
  restructuring (1962–67) 119–121
  state involvement/decline (1980–2000) 123–128, 130

suppliers 127–128
technological dynamism 126–128
technology policy 115, 119, 120, 124, 129–131
and TNCs 115, 121, 129–130
trade barriers/tariffs 124–125, 130
Brazilian civil aircraft manufacturing 5, 78–79, *91*
changing features 85–93
collaborative agreements 84–85
competition 79, 89–90
economic performance 91, *92*
export markets 88, 89, 91
general characteristics 89–93
nationalisation programme 88–89
post-privatisation (1995–2002) 95–98, 101–102
pre-1969 period 86–87
production and knowledge systems 82–85, *83*, 93–95, *96*, 102
supply chain structure 82, 84, 87–88, 100
see also BASI (Brazilian civil aircraft system of innovation); Embraer (Empresa Brasileira de Aeronautica S.A.)
Brazilian Institute for Development and Coordination of the Aerospace Industry (IFI) 88
Brazilian Ministry of Aeronautics 87
Brazilian Ministry of Defence 86, 88, 91, 94
Brazilian Technological Institute of Aeronautics (ITA) 86–88, 95
Breschi, S. 182
Bresnahan, T. 179
British Aerospace (BAe) 89, 90
Brusco, S. 182

Canadair (Bombardier Aerospace) 89, 90
Capability Maturity Model (CMM) 188, 215
Carnegie Mellon Research Institute for Information Systems Engineering 196, 215
case studies
Bangalore software cluster 180, 183–198
Brazilian automotive industry 108, 115–128
Brazilian civil aircraft industry 77–102
Cebu (Philippines) furniture industry 8, 159, 164–173
Punjabi farm equipment manufacturing 137, 149–152

Singapore's growth process 40–44
Cassiolato, J. E. 80
caste system 202–203, 205
Brahmins 10, 202, 205, 206, 209–210, 211–212, 217–221
Dravidian culture 211–212
education/learning emphasis 212–214, *213*, 220–221
and Indian economy 207–208
industrialisation and entrepreneurship 210–211, 220
jatis and varnas 206–207, 209, 210, 221
Ksatriyas 212, 218
Parsis, Jains and Sikhs 209, 220
and political stability 212
Sankritisation 210, 211–212
untouchables 206
upper and lower groups 209–211
Vaishas 10, 202, 209, 211, 212, 218, 220, 221
catch-up
in Asian tiger economies 203
in developing countries 1, 24–26, 48, 179–180, 198
Singaporean 61
Cebu Furniture Industry Foundation 169
Cebu (Philippines) furniture industry 8, 159, 164–165
cooperation and competition 167–168
effects of September 11 events 166–167
knowledge protection and transmission 170–173
localised learning capacity 165–168, 173–174
skilled workers and learning 168–173
subcontractors and exporters 166–168, 170–172, 173
training and skills development 170–171
workforce survey 169–170, *171*
working conditions 169
Central Machine Tools Institute (India) 184
Cessna (aircraft manufacturer) 89, 90
Chalam, K. S. 212
Chang, H. J. 111
chemical products
exports 19, 32
in Singapore clusters 41
China 89
exports 16
furniture industry 8, 165
and "Indian model" 214

China (cont.)
  innovation/patenting activity 36, 37
  RCA indices 29
  and regionalisation 58
  technological competitiveness 39
Chrysler Corporation 118, 122
civil aircraft industry
  competition and subsidies 90
  international crisis 92, 93
  recession 89
clusters/clustering 2, 135–137
  agglomeration advantages 139–140, *141*
  artisan-based 161, 163, 169
  and catch-up strategies 24–26, 48, 179–180, 198
  common culture benefits 167–168
  cost savings and economies 140–142
  definition/typology 135, 182
  in developing countries 161–163, 178–180
  and dynamic growth 136
  European 167
  GLOBE project 211
  industrial advantages 137
  innovative capabilities 198
  and institution intervention 182
  and knowledge accumulation processes 137, 139
  and knowledge-intensive production 180–183
  literature and research studies 136, 137, 161, 180, 181
  negative effects 144, 148
  regional 7–8, 135, 178–180
  Singaporean 40–41, 58–60
  and SMEs 135–136
  survivalist 148
  and technology-intensive production 179
  see also Bangalore software cluster; small-enterprise clusters; software technology parks; software technology parks (STPs)
Coelho, F. P. 89
collective efficiency (CE)
  active/passive distinction 139–140, 145
  and economic growth 136–137, 152
  meso-level approaches 137–144
competition/competitiveness
  indicators 31–32, 37–38
  and informatisation 178
  and learning processes 159–160
  regional 146, 160
  and SMEs 135, 136–137
  and state intervention 110
computer and office equipment industry 18, 19, 32, 33
COMTRADE (UN Commodity Trade Statistics Database) 13, 15
  and high tech definition 17, 18
Conceição, P. 160
consumer goods sector 137
CTA *see* Brazilian Aerospace Technical Centre
culture
  and caste and ethnicity variables 202, 204, 205
  definition 203, 204, 205
  and economic development 203–205
  role in cluster dynamics 2, 167–168

Dagnino, R. 86, 87
Dahlman, D. 81
Davis, L. 18
De Havilland aircraft company 89
Denmark 21
dependency theory 114
Deshpande, A. 220
developed countries
  high tech sector 20, 22, 24–26, 28, 31, *32*, 35
  infrastructure 39
  national orientation 39
  patent activity *36*
  product specialisation/capacity 32–34, 39
  research institutes 114
  technological development/competitiveness 13, 37–40, *39*
developing countries
  artisan-based production 161, 163, 169
  and dependency theory 114
  high tech exports 24–26, *27*, 43
  learning and knowledge exchange 161–163
  national orientation 31, 37, 39
  patenting records 34–37, 39–40
  product specialisation/capacity 32–34, 39
  R&D capability 37
  socio-economic infrastructure 31, 39
  as subcontractors to MNCs 40
  technological competitiveness 37–40, *39*
  technological development and state intervention 107–109
  technological infrastructure 31–32, 39

development process
  regional 162
  related stages model 58
  strategies 48–49
development trajectories 8, 77, 80, 165
  and clustering factors 180
  high-/low-road 179, 181, 184, 189, 191, 196, 198
  Singaporean 50, 51–56
Dicken, P. 114
DiMaggio, P. 204
Donângelo, A. 89
Dornier/Fairchild 89, 90
Dossani, R. 212
Draper International, TDICI 193
Dumont, L. 211
Dumont, Louis 208
Dumont, Santos 86
dynamic economics 77
dynamic growth
  and clusters/clustering 136, 178–180
  and technological knowledge 109–110

East Asian NICs 70, 110
  exports of manufactured goods 15
  golden policy prescriptions 113
  and private sector 110
  and South Asian societies 205
  and technology policies 113–115
economic change, and learning processes 159
Economic Development Board, Singapore (EDB) 57, 61
economic growth/development
  and cross-cultural comparisons 204
  effective policy making 153
  literature 205
  in low-income regions 178–180
  and market failures 110
  and regional agglomeration 7
  and the state 112
  and technological development 77, 136–137
economic liberalisation and deregulation
  and automotive industry 124–126, 130
  in Brazil 79, 81, 90, 95
  effects on local economies 135
economics/economists
  literature 214
  and technology policies 109, 111

EDB see Economic Development Board (Singapore)
education and training
  and policy-making framework 145–146
  programmes 49
electrical equipment exports 19, 32
electronics/telecommunications industry
  as high tech products 19, 32
  Singapore 41, 43, *44*
  specialisation *33*
embedded autonomy 111–112
Embraer (Empresa Brasileira de Aeronautica S.A.) 5, 78, 81, 85–86, 94, 100
  economic performance 2, 91–92
  financial crisis 95
  founding 87
  privatisation 89, 90–91, 92
  risk-sharing agreements 97–98, 100
  supply chain 94–95
  see also BASI
engineering systems, in Singapore clusters 41
entrepreneurial states 110–111
entrepreneurs, and subcontractors 8
entrepreneurship
  Indian 202
  Singapore 57
environmental technology, in Singapore clusters 41
European Commission (EC) 195
European Union (EU) 195
Evans, P. 111, 113, 117
evolutionary economics literature 77
Executive Group for the Automotive Industry see GEIA (Grupo Executivo da Industria Automobilística (Brazil))
exports
  of manufactured goods 14–16
  net export ratios 28–30
  RCA indices 27, *29*
  see also high technology sector
express package transportation, global patenting trends 36

Fairchild/Dornier 89, 90
farm equipment industry (Pakistan) 7, 137, 149–152
  CE and TC perspectives 149–150
  development and growth 149–152
  incentives for innovation 150–151

farm equipment industry (Pakistan) (cont.)
  networking and collaboration 149–150
  science and technology infrastructure 150–151
  training programmes and institutes 151
FDI *see* foreign direct investment
Fiscal Incentives for Exports in Brazil *see* BEFIEX
Fokker (aircraft manufacturer) 89, 90
Ford Motors 118, 119, 122
foreign direct investment (FDI) 49
  in Brazil 127–128
  Singapore 42, 49–50, 56
France, and high tech exports 26
Frederking, L. 204
Frischtak, D. 81, 87, 88
Frobel, F. 180
Fromhold-Eisebith, M. 179, 184, 186
furniture industry
  Cebu (the Philippines) 8, 159, 164–173
  China 8, 165
  Indonesia 8, 165

Galbraith, J. K. 208
Gambar, A. 179
GDP (gross domestic product)
  Brazil 5, 91, 92, 122
  Philippines 8, 165
  Singapore 42, 51, 52, 53, 54, 63, 64
GEIA (Grupo Executivo da Industria Automobilística, Brazil) 116, 117
Gelblum, T. 209
General Electric 5
General Motors (GM) 118, 119, 122
GERD (gross expenditure on R&D) 41, 42, 63, 64
Germany 21, 26, 36
Ghana 136
Ghani, E. 140
Gintic Institute of Manufacturing Technology (Singapore) 68
globalisation
  and automotive industry 123
  and developing countries 48, 71, 158
  and Singapore 6, 49, 50, 58, 67, 71
  technological 49, 50, 67, 69
GNP (gross national product)
  Brazil 78
  Singapore 51, 52, 53
Grupo Executivo da Industria Automobilística, Brazil *see* GEIA

Hari, P. 196
Harrison, L. 203
Hatter, V. L. 18
Hatzichronoglou, T. 17–18
health: global patenting trends 36
Heeks, R. 216
Heesterman, J. 208
Heinrichs, J. 180
Heitor, M. 160
Hewlett-Packard company 62, 184
high technology sector
  capacity 32
  catch up of developing countries 24–26
  and clustering 180–183
  competitiveness 26–28, 29, 31–32, 37–40
  definitions 17–20
  export concentrations and structure 23–24, 25, 26, 32, 35
  finance and investment 192–195
  Hatter-Davis definition 20
  innovative activity 31, 180–183
  leading exporters 21
  main export features 22–30
  net export ratios 28–30
  OECD list/definition 19–20
  RCA indices 27, 29
  specialisation 31, 32–34
  and statistical artefact hypothesis 31–40
  in world trade 3–5, 17–22
Hinduism 10, 206, 207–208
  and behaviour 208–211
  education system and professions 209–210
  karma, dharma, jati-dharma and svadharma 207–209, 210–211
  see also caste system
Holmstrom, M. 184
Hong Kong 21, 39, 63, 214
Humphrey, J. 162
Hungary 21
Huntington, S. 203
Hyderabad 215, 216, 222

I-flex Solutions (banking software) 186
IBM (International Business Machines) 68, 182
  Information Technology Programme for Office Workers (Singapore) 71
Ichimura, J. 89
ICT *see* information and communication technology

INDEX 235

IFI *see* Institute for Development and Coordination of the Aerospace Industry (Brazil)
IMF (International Monetary Fund) 114
India
  economic culture and development 203–205
  entrepreneurship 202, 210–211, 220
  export of manufactured products 16
  knowledge economy 203
  networks 185–186
  NRIs (non-resident Indians) 184, 189, 191, 221
  product development 185
  and regionalisation 58
  societal behaviour and beliefs 208–211, 216–217, *218*
  software services industry 15, 179, 182, 184, 185–186
  technological competitiveness 39
  trade/export market 9, 184, 185
  see also caste system; Hinduism; Indian ICT industry
Indian Department of Electronics 195
Indian ICT industry 10, 179, 182–183, 185, 202–203
  certified companies 188, 196, 215
  clusters 215–216
  cultural/ethnic influences 10, 216–222, *217*, *218*
  domination by South Indian Brahmins 221–222
  economic development 215, 216
  export performance 214, 215
  foreign investment 215
  innovative capabilities 190, 215
  interview data results 215–221
  overview 214–215
  regional distribution 219–221
  software development enterprises 214–215
  software technology parks (STPs) 214, 215, 216, 219
  state support 214–215
  university industry linkages 195–196, 215
  workforce 219–222
  see also Bangalore software cluster; caste system; Hinduism
Indian Institute of Information Technology (Bangalore) 195
Indian Institute of Science (IISc) 183, 184, 195
Indian Institutes of Technology (IITs) 189, 215, 216
Indonesia 16, 39
  and clustering 164
  furniture industry 8, 165
  and regionalisation 58
industrial policies 147
  Brazil 77, 80, 81, 130
  and network dynamics 80
  Singapore 40, 42–43, 50
industrialisation
  in Brazil 78, 115–116, 121–122
  in India 210–211, 220
  in less developed countries (LDCs) 109–110
industry – university relations
  Bangalore 95–96
  in Singapore 67–71
  United States 67–68
information and communication technology (ICT)
  in low-income regions 178–180
  "software bottleneck" 188–189
  software services industry 178–179
  see also Indian ICT industry
information technology (IT)
  global patenting trends 36
  R&D costs 181
  in Singapore 60
Infosys Technologies (banking software) 186–187
infrastructure
  investment 48
  socio-economic 31, 39
  technological 31–32, 39
  upgrading 145
innovation systems 31, 77
  benchmark models 62–63
  and collaboration 165
  definitions 80
  funding 181–182
  generation and diffusion 49
  and geographical differences 77, 80
  as interactive learning process 80
  and knowledge spillovers 160
  literature 77, 79–85, 205
  and networks 142
  policies 48
  and regional cultures 205–214

innovation systems (cont.)
  and stimulating interventions 146–147
  and technical change 79
  and technological diffusion 77, 80
  see also BASI
innovative capabilities
  in Bangalore software industry 190, 215
  of clusters 198
Institute of Bio-Informatics (Bangalore) 195
Institute for Development and Coordination of the Aerospace Industry, Brazil (IFI) 88, 89
Institute for New Technologies (INTECH) 20–22
institutions
  adaptations 48–49
  building 48, 147
  interventions in high tech clusters 182
  and networks 48
  "pockets of efficiency" 117, 130
  technology support 181
integrated circuits (ICs) production, Singapore 43, 44
intellectual property rights 34
inter-firm relations 158, 160, 167–168
interactive learning process 49, 80
international automotive industry
  economic crisis 121
  global patenting trends 36
  globalisation of 123
  Japanese involvement 122–123
  market liberalisation 124–126, 130
  oil crisis/recession 123
  restructuring process 124–125
  technological dynamism 108
  trade barriers/tariffs 124–125
  US/European competition 118–119
  see also Brazilian automotive industry
international environment
  and Brazil 113–115, 117, 122, 123
  and Singapore 71
International Labour Organization (ILO) 167
International Monetary Fund (IMF) 114
International Standards Organization (ISO) 219–220
  9000 (ISO 9000) certification 188, 215
international trade
  auto industry competition 118–119
  BEFIEX (Brazilian Fiscal Incentives for Exports) 121, 124
  intra-firm 114
  and Law of Similars 89, 116, 117
  Mercosur Agreement (1994) 125
  secret laws 34
  tariffs 125
  see also international trade performances
international trade performances
  classifying 13–14
  high technology products 3–5, 17–22
  of manufactured products 17–22
  net export ratios 28–30
  RCA indices 27, *29*
  and statistical artefact hypothesis 31–40
  see also exports
investments
  joint ventures 84, 179, 184
  risk-sharing agreements 84, 85, 97, 99, 100
  see also foreign direct investment (FDI)
Ireland 21, 24, 25
Israel 21
IT *see* information technology
ITA *see* Technological Institute of Aeronautics (Brazil)

Jamaica 21, 22
Japan
  automotive industry 122–123
  competitiveness scores 37, 39
  and high tech exports 21, 25
  innovation system 63
  investment in Singapore 54–55
  technological competitiveness 39
Japanese – Singapore Institute of Software Technology 69
Jenkins, R. 114
Jordan 21, 22

Katz, J. 80
Keeble, D. 167
Kenya 136
Knorringa, P. 164
knowledge accumulation
  and clusters/clustering 137
  effects on technological effort 142–144
  and learning processes 77–78
knowledge diffusion
  and clustering 181
  effects on labour market 169
  negative mechanisms 168
  and skilled labour 158–159, 168–173, 174

knowledge spillovers
  and agglomeration advantages *141*, 142–144
  and collective institution-building initiatives 147
  group classification 140–148
  negative effects 144
  and spatial proximity 160
knowledge systems (KS)
  in Brazilian aircraft manufacturing 81, 82–85
  diffusion of 110
  interaction/integration with production systems 84–85
knowledge-intensive industries 219
  and clustering 180–183
Korea
  export of manufactured products 16
  and high tech exports 21, 25, 26, 43
  and "Indian model" 214
  infrastructure 113
  innovation/patenting activity 36, 37, 38
  net export ratios 28, 29
Kravis, I. B. 17
Kreye, O. 180
Kubitschek de Oliveira, Juscelino 115, 116
Kubitschek government (Brazil) 117
Kumar, N. 184, 188

labour markets 189–191
  see also workforces
Lall, D. 207, 208, 211
Lall, S. 17, 22, 31, 42, 109
Lastres, H. M. M. 80
Lateef, A. 184
Latin America 81, 136
Law of Similars 89, 116, 117
Lawson, C. 160, 160–161
LDCs (less developed countries) 109
learning processes
  conceptualised 159–161
  and cultural heritage 163–164, 174
  current debate 159
  formal/informal 160–161, 163, 174
  localised 158, 159–164, 165, 168, 173–175
  protectionist behaviour 170, 174
  and skilled workers 163–164, 168–173
  in small enterprise 159–163, 173
  socially embedded 165–166, 168
LG Soft India 195
life sciences 41, 60

Lipsey, R. E. 17
Local Industries Upgrading Programme (Singapore) 42
low technology 17, 31, 164
low-income economies
  catch-up ability 1, 24–26, 48, 179–180, 198
  growth performances 196
Lundvall, B. A. 80

Malayan Federation 51, 56
Malaysia
  exports 16, 21, 25, 26, 28, 43
  patenting activity 36, 37, 39
  RCA indices 27, 29
  and regionalisation 58
  technological competitiveness 39
Malerba, F. 182
Malta 21
manufactured products 14–22
  exports 14–16
markets
  economies 17–22, 109
  failures 110
  governance 80
Massachusetts Institute of Technology (MIT) 67, 86
Mastek Software Systems 193
Menzies, H. 112–113
Mercosur (Southern Common Market Agreement, 1994) 6, 125
Mericle, S. K. 119
Mexico 16, 21, 29
Morocco 21, 22
Motorola 9
Motoyama, Y. 219
Mowery, D. 90
multinational corporations (MNCs) 9, 43, 49
  and Bangalore software cluster 179, 196
  role as strategic partners with governments 71
  in Singapore 40, 42, 43
Mumbai *see* Bombay

Nanyang Technological University (NTU), Singapore 68, 69
NASA (National Aeronautics & Space Administration (USA)) 94
National Aeronautical Laboratory (India) 184

National Bank for Economic Development (Banco Nacional de Desenvolvimento Econo mico, BNDE), Brazil 116
national orientation and competitiveness 7, 31, 39
National Science and Technology Board (NSTB), Singapore 40, 41
National System of Scientific and Technological Development (SNDCT), Brazil 108
National University of Singapore (NUS) 68, 69
National Venture Capital Fund for the Software and IT Industry (NFSIT), India 194–195
Neiva 87
net export ratios 28–30
Netherlands 21, 25
networks/networking
  and agglomeration advantages 145
  in Bangalore cluster 9
  and clusters 137, 142–144, 180
  institution 48
  and systems of innovation 77, 80
  and technological learning 153
New International Division of Labour proposition 180
New York Stock Exchange (NYSE) 90
Nicaragua 21, 22
NICs (newly industrialised countries) see East Asian NICs
non-electrical equipment exports 19, 32
North, D. 204
North India 212, 218, 219, 220
NSTB see National Science and Technology Board
NTU see Nanyang Technological University
NUS see National University of Singapore
NYSE see New York Stock Exchange

OECD (Organisation for Economic Cooperation and Development) 13, 17
  definition of high tech products 19–20
office equipment see computer and office equipment industry
Office of Technology Policy 36
Olson, M. 207
Oman, C. 114
Organisation for Economic Cooperation and Development see OECD

Padgett, T. 91, 93
Pakistan 7, 137
  Farm Machinery Institute, Islamabad 150–151
  Punjab region 149
  Punjab Small Industries Corporation 151
  see also farm equipment industry (Pakistan)
Paraguay 125
Parry, J. 207, 209
Parthasarathy, A. 184
patents/patenting: in developing countries 34–37, 49
Patibandla, M. 179, 184
Pavitt, K. 34
Perez-Aleman, P. 161
Petersen, B. 179, 184
Petrobrás (Brazil) 81
pharmaceutical industry 19, 32, 36
Philippines
  export of manufactured products 16
  furniture industry 8, 159, 164–173
  high tech exports 21, 25, 43
  innovation/patenting activity 37
  net export ratios 28, 29
Philips Corporation 62
  Industrial Engineering Programme, Singapore 71
Piper aircraft manufacturers (USA) 89, 94
policy making 112, 114, 144–153
political economy analysis
  approach to technology policies 112–113
  international environment 113–115
  key dimensions 112–115
  national historical and institutional characteristics 113
  new framework 112–115, 117, 119, 128–130
  relations between agents 113
political scientists 111–112
Porter, Michael 58–59
printed circuit board (PCB) industry 43, 44
private sector
  Brazil 110, 117
  and clustering 147
  in East Asian NICs 110
  Singapore 61, 62, 64, 69
  and software training 191
production systems (PS)
  in Brazilian aircraft manufacturing 82–85

INDEX 239

interaction/integration with knowledge systems 84–85
specialisation 31
Proença, D. J. 86, 87
protectionism in infant industries 109–110
Pyke, F. 162, 164

Quadros, R. A. 81
Quan, X. 219

RCA (revealed comparative advantage) indices 27
regional agglomeration/clusters
 literature 7
 and networking 181
 of SMEs 135–136
Regional Engineering Colleges (India) 189
regionalisation
 Mercosur Agreement (1994) 125
 and Singapore 58
Relatório 118
research and development (R&D) 49
 and agglomeration advantages 142–144
 embodied 18
 gross expenditure on (GERD) 41, 42, 63, 64
 and industry university relations 67–71
 intensity definition 17
 and local knowledge agglomeration 61, 62
 in Singapore clusters 41–42, 63, 64
research scientists and engineers (RSEs) 63–64
revealed comparative advantage indices *see* RCA indices
revealed technology advantage indexes 34
risk-sharing agreements 84–85, 97, 99, 100
Rodríguez-Pose, A. 127
Rolls Royce 5
Rosenberg, N. 90
Ruffles, P. 82, 84, 85

Saab (aircraft manufacturer) 89, 90
Sandee, H. 163
Saudi Arabia 21
Saxenian, A. 179, 216, 219
Schmitz, H. 162, 166
Schumpeter, J. 80
science and technology (S&T) sector 4, 13, 39, 60–63
scientific instruments exports 19, 32

SEI-CMM (Software Engineering Institute's Capability Maturity Model, USA) 188
Sen, A. 205
Senegal 21, 22
Sengenberger, W. 162, 164
Shapiro, H. 117, 122, 123
Shastri, V. 216
Silicon Valley 9, 67, 179, 195, 206, 214
Simputer Trust 196
Singapore
 academic and research centres 67–71
 chemical industry 53
 clusters/clustering 40–41, 58–60
 competitiveness 37, 39, 56–60
 development strategies 49–50, 51–60, 71–72
 economy/GDP 50, 51, 52, 53–54, 56–60, 61–62, 63–67, 69, 71
 education/sustained learning 41–42, 55–56, 59, 61, 67–71
 electronics industry 43, *44*, 53, 55, 66
 engineering sciences 64, 66, 69
 export growth case study 40–44
 foreign investment 42, 53, 54–55, 56, 60, 66, 71
 GERD/GDP ratio 63–64
 and globalisation 6, 49, 50, 58, 67, 69, 71
 growth and development trajectory 51–56
 high tech exports 21, 25, 26, 43
 industrial policies 40, 42–43, 50, 55, 58, 59–60, 64–65
 infrastructure 48, 51, 54
 innovation policies 40–41, 50, 59, 60–71
 institution building/networks 60, 67, 71–72
 "intelligent island" development strategy 50, 60, 66
 IT-related services sector 66
 and Malayan Federation 51, 56
 manufacturing sector 16, 51, 53, 56, 60–61, 66
 and MNCs 40, 42, 43, 53–54, 58, 59, 60, 65, 71
 National Science and Technology Plan 2000 40
 net export ratios 28
 patenting activity 36, 37, 38, 63
 R&D activities/investment 41–42, 59, 60–71
 RCA indices 27, 29
 and regionalisation 58

Singapore (cont.)
    science and technology sector 4, 39, 60–63
    service sector 53–54, 56, 61
    taxation/fiscal schemes 59, 62
    telecommunications sector 57
    wage policies 57–58
    workforce/training programmes 69, 70–71
Singapore Advanced Materials Research Centre 68
Singapore Economic Development Board (EDB) 57, 59, 61
Singapore Institute for Microelectronics 68
Singapore Institute for Molecular and Cell Biology 68
Singapore Institute for Systems Science 68
Singapore Institute of Technical Education 70–71
Singapore National Science and Technology Board (NSTB) 61, 62
Singapore National Wages Council 53, 56
Singapore Standards, Productivity and Innovation Board (SPRING) 70
small-enterprise clusters
    buyer-driven value chains 161–162
    and horizontal and vertical inter-firm relations 158, 167–168
    and internal hierarchies 162
    and localised learning 158, 159–163
SMEs (small-/medium-sized enterprises) 135
    and clustering dynamics 135–137
    competitiveness 137, 152, 153
    European 167
    in Singapore 42, 69, 72
socio-cultural institutions, and network dynamics 80
Soete, L. 34
software development industry 9–10
    export trade 15
    innovations 188–189
    and internet 189
    and labour markets 189–191
    in low-income regions 178–180
    programming and codification skills 219, 220
    recruitment and training 189–190
Software Engineering Institute (Bangalore) 195

Software Engineering Institute (USA), SEI-CMM (Capability Maturity Model) 188
software process improvement network (SPIN) 188
Software, Services, Support and Education Centre Limited (3SE), India 195
software technology parks (STPs) 147, 192, 214, 215, 216, 219
Sony Corporation 68
South Asia societies 205
South India 10, 202, 221
    culture and ethnicity 211–214, 217–222, *217*, *218*
    economic culture 205–214
    education and learning 221–222
    investments 219
    networks 222
    technical and engineering colleges 212–214, *213*
    see also Bangalore software cluster; caste system; Hinduism
South Korea 39, 111, 117, 179
Southeast Asia 54–55
Southern Common Market Agreement *see* Mercosur
Standard Industrial Classification (SIC) 18
state intervention
    and entrepreneurialism 110–111
    and international competitiveness 110
    and market failures 109
    and technological development in developing countries 107–109
states
    and embedded autonomy 111–112
    entrepreneurial 110–111
    military regimes 111, 121
    and protectionism 109–110
    and technology policies 107–109, 111
statistical artefact hypothesis verification 31–40
Stewart, F. 140
Storper, M. 164
STPs *see* software technology parks
supply chains 87
    structure and tiers 82, 84
Sweden 21
Switzerland 21
systems of innovation (SI) *see* innovation systems

INDEX   241

Taiwan 36, 39, 113, 179
TDICI *see* Technology Development and Information Company of India
technological capability (TC)
  and agglomeration advantages 139–140, *141*, 144
  development indicators 13
  and geographical agglomeration 147
  high/low 31
  and innovation stimulation 153
  integration with collective efficiency (CE) 137–144, *138*
  literature 136–137, 138
  micro-level approach 137–144
  origins of term 138
  and patenting activity 34, 39
  and regional clustering 7, 136
technological collaboration 85, 138
technological development
  and economic growth 77
  and innovation systems 78
  and multinationals 179
  and state intervention in developing countries 107–109
  in traditional communities 148
Technological Institute of Aeronautics, Brazil (ITA) 86–87, 95
technological knowledge/learning
  classification 139–140
  and clusters 153
  and dynamic economies 109–110
Technology Development and Information Company of India (TDICI) 193
technology policies
  contrasting approaches *114*
  economists' contributions 109–111
  formation and evolution 108
  influences 131
  literature 108
  political economy framework 112–115, 117, 119, 128–130
  research 107
  review of theories 109–112
  and state autonomy 111
  state protectionism and coordination 109–110
  see also Brazilian automotive industry
Technology Policy Assessment Center (TPAC), Georgia USA 31
technology-intensive production 179

telecommunications *see* electronics/telecommunications industry
Texas Instruments (TI) 9, 183–184
Thailand
  export of manufactured products 16
  high tech exports 21, 25, 43
  innovation/patenting activity 37
  net export ratios 28
  RCA indices 29
Thapar, R. 207
TNC *see* transnational corporations
Toyota 122
trade *see* international trade
transnational corporations (TNCs)
  and Brazilian automotive industry 6, 115, 118, 121, 129–130
  intra-firm trade 114
  involvement in Bangalore cluster 184, 222
  and technology policies in developing countries 114
Trinidad and Tobago 21, 22
Tschang, T. 216
Tyler, W. G. 122

UN Commodity Trade Statistics Database *see* COMTRADE
United Kingdom (UK)
  and Brazilian aircraft industry 89
  and high tech exports 21, 25, 26
  patenting activity 36
United Nations Industrial Development Organisation (UNIDO) 20
  Yearbook 17
United States Patents and Trademarks Office (USPTO) 37
United States (USA)
  auto industry and foreign competition 118–119
  and Brazilian aircraft industry 86, 89, 91
  competitiveness scores 37, 39
  Department of Commerce 18
  global patents classes and grants 34–37, 38, 63
  and high tech exports 21, 25, 26
  and Indian labour markets 190
  industry – university relations 67–68
  September 2001 attacks 92, 93, 166–167
  technological competitiveness 39
Uruguay 125

USPTO *see* United States Patents and
 Trademarks Office

value chains 137
 and agglomeration advantages 144
 Bangalore software cluster 179, 198
 buyer-driven 161–162
 and dynamic learning 153
 and knowledge exchange 167
venture capital (VC) 192–195
Viotti, E. B. 80

Walden International 193
Washington Consensus 124, 130
Weber, Max 205, 208, 211
Wipro Technologies 193, 215
workforces
 artisan-based 161, 163, 169

Cebu survey 169–170, *171*
 in high technology sectors 181–182
 inter-firm division of labour 181
 and knowledge diffusion 158–159, 161, 163
 labour markets 189–191
 mobility 9–10, 145, 161, 163
 piracy of 168
 skilled 163–164, 168–173
 skills development and training 163–164
 work practice improvements 146
World Bank (WB) 18
 and high tech products 20–22
World Development journal 161
World Technology Evaluation Center 43
world trade *see* international trade
 performances
World Trade Organization (WTO) 90

Catalogue Request

Name: _____

Address: _____

Tel: _____

Fax: _____

E-mail: _____

To receive a catalogue of UNU Press publications kindly photocopy this form and send or fax it back to us with your details. You can also e-mail us this information. Please put "Mailing List" in the subject line.

United Nations University Press

53-70, Jingumae 5-chome
Shibuya-ku, Tokyo 150-8925, Japan
Tel: +81-3-3499-2811   Fax: +81-3-3406-7345
E-mail: sales@hq.unu.edu   http://www.unu.edu